Basic Electronics for Scientists and Engineers

Ideal for a one-semester course, this concise textbook covers basic electronics for undergraduate students in science and engineering.

Beginning with basics of general circuit laws and resistor circuits to ease students into the subject, the textbook then covers a wide range of topics, from passive circuits through to semiconductor-based analog circuits and basic digital circuits. Using a balance of thorough analysis and insight, readers are shown how to work with electronic circuits and apply the techniques they have learnt. The textbook's structure makes it useful as a self-study introduction to the subject. All mathematics is kept to a suitable level, and there are several exercises throughout the book. Solutions for instructors, together with eight laboratory exercises that parallel the text, are available online at www.cambridge.org/Eggleston.

Dennis L. Eggleston is Professor of Physics at Occidental College, Los Angeles, where he teaches undergraduate courses and labs at all levels (including the course on which this textbook is based). He has also established an active research program in plasma physics and, together with his undergraduate assistants, he has designed and constructed three plasma devices which form the basis for the research program.

Basic Electronics for Scientists and Engineers

Dennis L. Eggleston

Occidental College, Los Angeles

CAMBRIDGE
UNIVERSITY PRESS

University Printing House, Cambridge CB2 8BS, United Kingdom

One Liberty Plaza, 20th Floor, New York, NY 10006, USA

477 Williamstown Road, Port Melbourne, VIC 3207, Australia

314-321, 3rd Floor, Plot 3, Splendor Forum, Jasola District Centre, New Delhi - 110025, India

103 Penang Road, #05-06/07, Visioncrest Commercial, Singapore 238467

Cambridge University Press is part of the University of Cambridge.

It furthers the University's mission by disseminating knowledge in the pursuit of education, learning and research at the highest international levels of excellence.

www.cambridge.org
Information on this title: www.cambridge.org/9780521154307

First published 2011
9th printing 2019

A catalogue record for this publication is available from the British Library

Library of Congress Cataloging in Publication data
Eggleston, Dennis L. (Dennis Lee), 1953-
Basic Electronics for Scientists and Engineers / by Dennis L. Eggleston.
p.   cm
Includes bibliographical references and index.
ISBN 978-0-521-76970-9 (Hardback) – ISBN 978-0-521-15430-7 (Paperback)   1. Electronics.
I. Title.
TK7816.E35 2011
621.381–dc22

                                                                    2010050327

ISBN 978-0-521-76970-9 Hardback
ISBN 978-0-521-15430-7 Paperback

Additional resources for this publication at www.cambridge.org/Eggleston

To my wife Lynne

Contents

Preface   *page* xi

1 Basic concepts and resistor circuits   **1**
    1.1 Basics   1
    1.2 Resistors   4
    1.3 AC signals   19
         Exercises   23
         Further reading   26

2 AC circuits   **27**
    2.1 Introduction   27
    2.2 Capacitors   27
    2.3 Inductors   29
    2.4 RC circuits   30
    2.5 Response to a sine wave   37
    2.6 Using complex numbers in electronics   43
    2.7 Using the complex exponential method for a switching problem   54
    2.8 Fourier analysis   58
    2.9 Transformers   61
         Exercises   65
         Further reading   67

3 Band theory and diode circuits   **68**
    3.1 The band theory of solids   68
    3.2 Diode circuits   80
         Exercises   101
         Further reading   103

4 Bipolar junction transistors   **104**
    4.1 Introduction   104
    4.2 Bipolar transistor fundamentals   104

4.3	DC and switching applications	108
4.4	Amplifiers	110
	Exercises	131
	Further reading	132

5 Field-effect transistors   **133**

5.1	Introduction	133
5.2	Field-effect transistor fundamentals	134
5.3	DC and switching applications	140
5.4	Amplifiers	141
	Exercises	150
	Further reading	151

6 Operational amplifiers   **152**

6.1	Introduction	152
6.2	Non-linear applications I	153
6.3	Linear applications	154
6.4	Practical considerations for real op-amps	159
6.5	Non-linear applications II	165
	Exercises	168
	Further reading	170

7 Oscillators   **171**

7.1	Introduction	171
7.2	Relaxation oscillators	171
7.3	Sinusoidal oscillators	185
7.4	Oscillator application: EM communications	193
	Exercises	198
	Further reading	199

8 Digital circuits and devices   **200**

8.1	Introduction	200
8.2	Binary numbers	200
8.3	Representing binary numbers in a circuit	202
8.4	Logic gates	204
8.5	Implementing logical functions	206
8.6	Boolean algebra	208
8.7	Making logic gates	211

8.8   Adders   213
8.9   Information registers   216
8.10 Counters   220
8.11 Displays and decoders   223
8.12 Shift registers   224
8.13 Digital to analog converters   227
8.14 Analog to digital converters   228
8.15 Multiplexers and demultiplexers   229
8.16 Memory chips   232
      Exercises   234
      Further reading   235

Appendix A: Selected answers to exercises   236
Appendix B: Solving a set of linear algebraic equations 238
Appendix C: Inductively coupled circuits   241
References   245
Index   247

Preface

A professor of mine once opined that the best working experimentalists tended to have a good grasp of basic electronics. Experimental data often come in the form of electronic signals, and one needs to understand how to acquire and manipulate such signals properly. Indeed, in graduate school, everyone had a story about a budding scientist who got very excited about some new result, only to later discover that the result was just an artifact of the electronics they were using (or misusing!). In addition, most research labs these days have at least a few homemade circuits, often because the desired electronic function is either not available commercially or is prohibitively expensive. Other anecdotes could be added, but these suffice to illustrate the utility of understanding basic electronics for the working scientist.

On the other hand, the sheer volume of information on electronics makes learning the subject a daunting task. Electronics is a multi-hundred billion dollar a year industry, and new products of ever-increasing specialization are developed regularly. Some introductory electronics texts are longer than introductory physics texts, and the print catalog for one national electronic parts distributor exceeds two thousand pages (with tiny fonts!).

Finally, the undergraduate curriculum for most science and engineering majors (excepting, of course, electrical engineering) does not have much space for the study of electronics. For many science students, formal study of electronics is limited to the coverage of voltage, current, and passive components (resistors, capacitors, and inductors) in introductory physics. A dedicated course in electronics, if it exists, is usually limited to one semester.

This text grew out of my attempts to deal with this three-fold challenge. It is based on my notes for a one-semester course on electronics I have taught for many years in the Physics Department of Occidental College. The students in the course are typically sophomore, junior, or senior students majoring in physics or pre-engineering, with some from the other sciences and mathematics. The students have usually had at least two introductory physics courses and two semesters of calculus.

The primary challenge of such a course is to select the topics to include. My choices for this text have been guided by several principles: I wanted the text to be a rigorous, self-contained, one-semester introduction to basic analog and digital electronics. It should start with basic concepts and at least touch upon the major topics. I also let the choice of material be guided by those topics I thought were fundamental or have found useful during my career as a researcher in experimental plasma physics. Finally,

I wanted the text to emphasize learning how to work with electronics through analysis rather than copying examples.

Chapters 1 and 2 start with basic concepts and cover the three passive components. Key concepts such as Thevenin's theorem, time- and frequency-domain analysis, and complex impedances are introduced. Chapter 3 uses the band theory of solids to explain semiconductor diode operation and shows how the diode and its cousins can be used in circuits. The use of the load line to solve the transcendental equations arising from the diode's non-linear I–V characteristic is introduced, as well as common approximation techniques. The fundamentals of power supply construction are also introduced in this chapter.

Bipolar junction transistors and field-effect transistors are covered in Chapters 4 and 5. Basic switching and amplifier circuits are analyzed and transistor AC equivalents are used to derive the voltage and current gain as well as the input and output impedance of the amplifiers. A discussion of feedback in Chapter 4 leads into the study of operational amplifiers in Chapter 6. Linear and non-linear circuits are analyzed and the limitations of real op-amps detailed.

Several examples of relaxation and sinusoidal oscillators are studied in Chapter 7, with time-domain analysis used for the former and frequency-domain analysis used for the latter. Amplitude- and frequency-modulation are introduced as oscillator applications. Finally, a number of basic digital circuits and devices are discussed in Chapter 8. These include the logic gates, flip-flops, counters, shift-registers, A/D and D/A converters, multiplexers, and memory chips. Although the digital universe is much larger than this (and expanding!), these seem sufficient to give a laboratory scientist a working knowledge of this universe and lay the foundation for further study.

Exercises are given at the end of each chapter along with texts for further study. I recommend doing all of the exercises. While simple plug-in problems are avoided, I have found that most students will rise to the challenge of applying the techniques studied in the text to non-trivial problems. Answers to some of the problems are given in Appendix A, and a solution manual is available to instructors.

At Occidental this course is accompanied by a laboratory, and I enthusiastically recommend such a structure. In addition to teaching a variety of laboratory skills, an instructional laboratory in electronics allows the student to connect the analytical approach of the text to the real world. A set of laboratory exercises that I have used is available from the publisher.

The original manuscript was typeset using LaTeX and the figures constructed using *PSTricks: Postscript macros for Generic TeX* by Timothy Van Zandt and *M4 Macros for Electric Circuit Diagrams in Latex Documents* by Dwight Aplevich. I am indebted to the makers of these products and would not have attempted this project without them.

Dennis L. Eggleston   Los Angeles, California, USA

"*Basic Electronics for Scientists and Engineers* by Dennis Eggleston is an example of how the most important material in the introduction to electronics can be presented within a one-semester time frame. The text is written in a nice logical sequence and is beneficial for students majoring in all areas of the Natural Science. In addition, many examples and detailed introduction of all equations allows this course to be taught to students of different background – sophomores, juniors, and seniors. Overall, the effort of the author is thrilling and, definitely, this text will be popular among many instructors and students."

Anatoliy Glushchenko, Department of Physics and Energy Science, University of Colorado at Colorado Springs

"This text is an excellent choice for undergraduates majoring in physics. It covers the basics, running from passive components through diodes, transistors and op-amps to digital electronics. This makes it self-contained and a one-stop reference for the student. A brief treatment of the semiconductor physics of silicon devices provides a good basis for understanding the mathematical models of their behaviour and the end-of-chapter problems help with the learning process. The concise and sequential nature of the book makes it easier to teach (and study) from than the venerable but somewhat overwhelming Art of Electronics by Horowitz and Hill."

David Hanna, W C Macdonald Professor of Physics, McGill University

"I have been frustrated in the past by my inability to find a suitable book for a one-semester Electronics course that starts with analog and progresses to basic digital circuits. Most available books seem to be out of date or aimed at electrical engineers rather than scientists. Eggleston's book is exactly what I was looking for – a basic course ideal for science students needing a practical introduction to electronics. Written concisely and clearly, the book emphasizes many practical applications, but with sufficient theoretical explanation so that the results don't simply appear out of thin air."

Susan Lehman, Clare Boothe Luce Associate Professor and Chair of Physics, The College of Wooster

1 Basic concepts and resistor circuits

1.1 Basics

We start our study of electronics with definitions and the basic laws that apply to *all* circuits. This is followed by an introduction to our first circuit element – the resistor.

In electronics, we are interested in keeping track of two basic quantities: the *currents* and *voltages* in a circuit. If you can make these quantities behave like you want, you have succeeded.

Current measures the flow of charge past a point in the circuit. The units of current are thus coulombs per second or *amperes*, abbreviated as A. In this text we will use the symbol I or i for current.

As charges move in circuits, they undergo collisions with atoms and lose some of their energy. It thus takes some work to move charges around a circuit. The work per unit charge required to move some charge between two points is called the *voltage* between those points. (In physics, this work per unit charge is equivalent to the difference in electrostatic potential between the two points, so the term *potential difference* is sometimes used for voltage.) The units of voltage are thus joules per coulomb or *volts*, abbreviated V. In this text we will use the symbol V or v for voltage.

In a circuit, there are sources and sinks of energy. Some sources of energy (or voltage) include batteries (which convert chemical energy to electrical energy), generators (mechanical to electrical energy), solar cells (radiant to electrical energy), and power supplies and signal generators (electrical to electrical energy). All other electrical components are sinks of energy.

Let's see how this works. The simplest circuit will involve one voltage source and one sink, with connecting wires as shown in Fig. 1.1. By convention, we denote the two sides of the voltage source as + and −. A positive charge moving from the − side to the + side of the source gains energy. Thus we say that the voltage across the source is positive. When the circuit is complete, current flows out of the + side of the source, as shown. The voltage across the component is negative when we

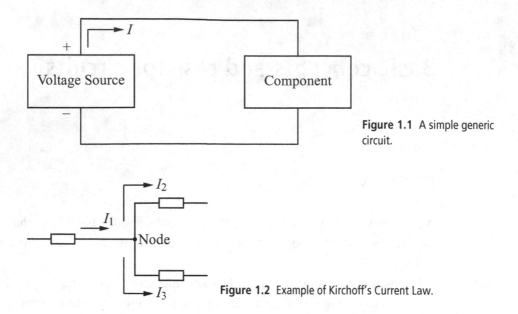

Figure 1.1 A simple generic circuit.

Figure 1.2 Example of Kirchoff's Current Law.

cross it in the direction of the current. We say there is a *voltage drop* across the component. Note that while we can speak of the current at any point in the circuit, the voltage is always between two points. It makes no sense to speak of the voltage at a point (remember, the voltage is a potential *difference*).

We can now write down some general rules about voltage and current.

1. The sum of the currents into a node (i.e. any point on the circuit) equals the sum of the currents flowing out of the node. This is Kirchoff's Current Law (KCL) and expresses conservation of charge. For example, in Fig. 1.2, $I_1 = I_2 + I_3$. If we use the sign convention that currents into a node are positive and currents out of a node are negative, then we can express this law in the compact form

$$\sum_k^{\text{node}} I_k = 0 \qquad (1.1)$$

where the sum is over all currents into or out of the node.

2. The sum of the voltages around any closed circuit is zero. This is Kirchoff's Voltage Law (KVL) and expresses conservation of energy. In equation form,

$$\sum_k^{\text{loop}} V_k = 0. \qquad (1.2)$$

Here we must use the convention that the voltage across a source is positive when we move across the source in the direction of the current and the voltage

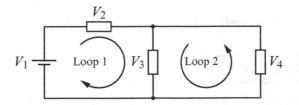

Figure 1.3 Example of Kirchoff's Voltage Law.

across a sink is negative when we move across the component in the direction of the current. If we traverse a source or sink in the direction opposite to the direction of the current, the signs are reversed. Figure 1.3 gives an example. Here we introduce the circuit symbol for an ideal battery, labeled with voltage V_1. The top of this symbol represents the positive side of the battery. The current (not shown) flows up out of the battery, through the component labeled V_2 and down through the components labeled V_3 and V_4. Looping around the left side of the circuit in the direction shown gives $V_1 - V_2 - V_3 = 0$ or $V_1 = V_2 + V_3$. Here we take V_2 and V_3 to be positive numbers and include the sign explicitly. Going around the right portion of the circuit as shown gives $-V_3 + V_4 = 0$ or $V_3 = V_4$. This last equality expresses the important result that components connected in parallel have the same voltage across them.

3. The power P provided or consumed by a circuit device is given by

$$P = VI \tag{1.3}$$

where V is the voltage across the device and I is the current through the device. This follows from the definitions:

$$VI = \left(\frac{\text{work}}{\text{charge}}\right)\left(\frac{\text{charge}}{\text{time}}\right) = \frac{\text{work}}{\text{time}} = \text{power}. \tag{1.4}$$

The units of power are thus joules per second or *watts*, abbreviated W. This law is of considerable practical importance since a key part of designing a circuit is to employ components with the proper power rating. A component with an insufficient power rating will quickly overheat and fail when the circuit is operated.

Finally, a word about prefixes and nomenclature. Some common prefixes and their meanings are shown in Table 1.1. As an example, recall that the unit *volts* is abbreviated as V, and *amperes* or *amps* is abbreviated as A. Thus 10^6 volts = 1 MV and 10^{-3} amps = 1 mA. Notice that case matters: 1 MA $\neq$ 1 mA.

Table 1.1 Some common prefixes used in electronics

Multiple	Prefix	Symbol
10^{12}	tera	T
10^{9}	giga	G
10^{6}	mega	M
10^{3}	kilo	k
10^{-3}	milli	m
10^{-6}	micro	μ
10^{-9}	nano	n
10^{-12}	pico	p
10^{-15}	femto	f

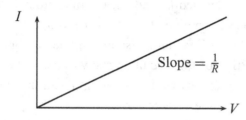

Figure 1.4 $I-V$ curve for a resistor.

1.2 Resistors

A common way to represent the behavior of a circuit device is the $I-V$ characteristic. This is a plot of the current I through the device as a function of applied voltage V across the device. Our first device, the resistor, has the simple linear $I-V$ characteristic shown in Fig. 1.4. This linear relationship is expressed by Ohm's Law:

$$V = IR. \tag{1.5}$$

The constant of proportionality, R, is called the *resistance* of the device and is equal to one over the slope of the $I-V$ characteristic. The units of resistance are *ohms*, abbreviated as Ω. Any device with a linear $I-V$ characteristic is called a resistor.

The resistance of the device depends only on its physical properties – its size and composition. More specifically:

$$R = \rho\frac{L}{A} \tag{1.6}$$

Table 1.2 The resistivity of some common electronic materials

Material	ρ (10^{-8} Ωm)
Silver	1.6
Copper	1.7
Nichrome	100
Carbon	3500

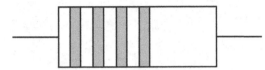

Figure 1.5 Value and tolerance bands on a resistor.

where ρ is the resistivity, L is the length, and A is the cross-sectional area of the material. The resistivity of some representative materials is given in Table 1.2.

The interconnecting wires or circuit board paths are typically made of copper or some other low resistivity material, so for most cases their resistance can be ignored. If we want resistance in a circuit we will use a discrete device made of some high resistivity material (e.g., carbon). Such resistors are widely used and can be obtained in a variety of values and power ratings. The low power rating resistors typically used in circuits are marked with color coded bands that give the resistance and the tolerance (i.e., the uncertainty in the resistance value) as shown schematically in Fig. 1.5.

As shown in the figure, the bands are usually grouped toward one end of the resistor. The band closest to the end is read as the first digit of the value. The next band is the second digit, the next band is the multiplier, and the last band is the tolerance value. The values associated with the various colors are shown in Table 1.3. For example, a resistor code having colors red, violet, orange, and gold corresponds to a value of 27×10^3 $\Omega \pm 5\%$.

Resistors also come in variable forms. If the variable device has two leads, it is called a *rheostat*. The more common and versatile type with three leads is called a *potentiometer* or a "pot." Schematic symbols for resistors are shown in Fig. 1.6.

One must also select the proper power rating for a resistor. The power rating of common carbon resistors is indicated by the size of the device. Typical values are $\frac{1}{8}, \frac{1}{4}, \frac{1}{2}, 1$, and 2 watts.

Table 1.3 Standard color scheme for resistors

Color	Digit	Multiplier	Tolerance (%)
none			20
silver		0.01	10
gold		0.1	5
black	0	1	
brown	1	10	
red	2	100	2
orange	3	10^3	
yellow	4	10^4	
green	5	10^5	
blue	6	10^6	
violet	7	10^7	
gray	8		
white	9		

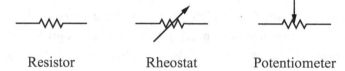

Resistor   Rheostat   Potentiometer

Figure 1.6 Schematic symbols for a fixed resistor and two types of variable resistors.

As noted in Eq. (1.3), the power consumed by a device is given by $P = VI$, but for resistors we also have the relation $V = IR$. Combining these we obtain two power relations specific to resistors:

$$P = I^2 R \qquad (1.7)$$

and

$$P = V^2/R. \qquad (1.8)$$

1.2.1 Equivalent circuit laws for resistors

It is common practice in electronics to replace a portion of a circuit with its functional equivalent. This often simplifies the circuit analysis for the remaining portion of the circuit. The following are some equivalent circuit laws for resistors.

1.2.1.1 Resistors in series

Components connected in series are connected in a head-to-tail fashion, thus forming a line or series of components. When forming equivalent circuits, any

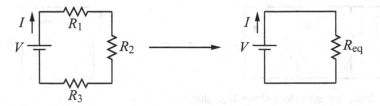

Figure 1.7 Equivalent circuit for resistors in series.

number of resistors in series may be replaced by a single equivalent resistor given by:

$$R_{eq} = \sum_i R_i \qquad (1.9)$$

where the sum is over all the resistors in series. To see this, consider the circuit shown in Fig. 1.7. We would like to replace the circuit on the left by the equivalent circuit on the right. The circuit on the right will be equivalent if the current supplied by the battery is the same.

By KCL, the current in each resistor is the same. Applying KVL around the circuit loop and Ohm's Law for the drop across the resistors, we obtain

$$V = IR_1 + IR_2 + IR_3$$
$$= I(R_1 + R_2 + R_3)$$
$$= IR_{eq} \qquad (1.10)$$

where

$$R_{eq} = R_1 + R_2 + R_3. \qquad (1.11)$$

This derivation can be extended to any number of resistors in series, hence Eq. (1.9).

1.2.1.2 Resistors in parallel

Components connected in parallel are connected in a head-to-head and tail-to-tail fashion. The components are often drawn in parallel lines, hence the name. When forming equivalent circuits, any number of resistors in parallel may be replaced by a single equivalent resistor given by:

$$\frac{1}{R_{eq}} = \sum_i \frac{1}{R_i} \qquad (1.12)$$

where the sum is over all the resistors in parallel. To see this, consider the circuit shown in Fig. 1.8. Again, we would like to replace the circuit on the left by the equivalent circuit on the right.

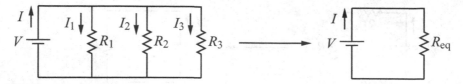

Figure 1.8 Equivalent circuit for resistors in parallel.

First, note that KCL requires

$$I = I_1 + I_2 + I_3. \tag{1.13}$$

Since the resistors are connected in parallel, the voltage across each one is the same, and, by KVL is equal to the battery voltage: $V = I_1R_1$, $V = I_2R_2$, $V = I_3R_3$. Solving these for the three currents and substituting in Eq. (1.13) gives

$$I = \frac{V}{R_1} + \frac{V}{R_2} + \frac{V}{R_3} = V\left(\frac{1}{R_1} + \frac{1}{R_2} + \frac{1}{R_3}\right) = \frac{V}{R_{eq}} \tag{1.14}$$

where

$$\frac{1}{R_{eq}} = \frac{1}{R_1} + \frac{1}{R_2} + \frac{1}{R_3}. \tag{1.15}$$

Again, this derivation can be extended to any number of resistors in parallel, hence Eq. (1.12).

A frequent task is to analyze two resistors in parallel. Of course, for this special case of Eq. (1.12) we get $\frac{1}{R_{eq}} = \frac{1}{R_1} + \frac{1}{R_2}$. It is often more illuminating to write this as an equation for R_{eq} rather than $\frac{1}{R_{eq}}$. After some algebra, we get

$$R_{eq} = \frac{R_1 R_2}{R_1 + R_2}. \tag{1.16}$$

This special case is worth memorizing.

Example For the circuit shown in Fig. 1.9, how much current flows through the 20 kΩ resistor? What must its power rating be?

Solution As we will see, there is more than one way to solve this problem. Here we use a method that relies on basic electronics reasoning and our resistor equivalent circuit laws. We want the current through the 20 kΩ resistor. If we knew the voltage across this resistor (call this voltage V_{20k}), we could then get the current from Ohm's Law. In order to get the voltage across the 20 kΩ resistor, we need the voltage across the 10 kΩ resistor since, by KVL, $V_{20k} = 130 - V_{10k}$. In order to get the voltage across the 10 kΩ resistor, we need to know the current through

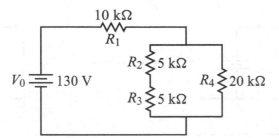

Figure 1.9 Example resistor circuit.

it, which is the same as the current supplied by the battery. Thus, if we can get the current supplied by the battery we can solve the problem. To get the battery current, we combine all our resistors into one equivalent resistor. The implementation of this strategy goes as follows.

1. Combine the two 5 kΩ series resistors into a 10 kΩ resistor.
2. This 10 kΩ resistor is then in parallel with the 20 kΩ resistor. Combining these we get (using Eq. (1.16))

$$R_{eq} = \frac{R_1 R_2}{R_1 + R_2} = \frac{(10 \text{ k}\Omega)(20 \text{ k}\Omega)}{10 \text{ k}\Omega + 20 \text{ k}\Omega} = 6.67 \text{ k}\Omega. \tag{1.17}$$

3. This 6.67 kΩ resistor is then in series with a 10 kΩ resistor, giving a total equivalent circuit resistance $R_{eq} = 16.67$ kΩ.
4. The current supplied by the battery is then

$$I = \frac{V_0}{R_{eq}} = \frac{130 \text{ V}}{16.67 \times 10^3 \text{ }\Omega} = 7.8 \times 10^{-3} \text{ A} = 7.8 \text{ mA}. \tag{1.18}$$

5. KVL then gives 130 V − (7.8 mA)(10 kΩ) − V_{20k} = 0. Solving this gives $V_{20k} = 52$ V.
6. Ohm's Law then gives $I_{20k} = \frac{52 \text{ V}}{20 \text{ k}\Omega} = 2.6$ mA, which is the solution to the first part of our problem. As a check, it is comforting to note that this current is less than the total battery current, as it must be. The remainder goes through the two 5 kΩ resistors.
7. The power consumed by the 20 kΩ resistor is $P = I^2 R = (2.6 \times 10^{-3} \text{ A})^2 (2 \times 10^4 \text{ }\Omega) = 0.135$ W. This is too much for a $\frac{1}{8}$ W resistor, so we must use at least a $\frac{1}{4}$ W resistor.

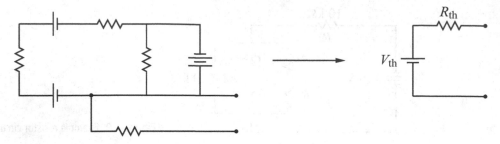

Figure 1.10 Representation of Thevenin's theorem.

1.2.1.3 Thevenin's theorem and Norton's theorem

The third of our equivalent circuit laws, Thevenin's theorem, is a more general result that actually includes the first two laws as special cases. The theorem states that any two-terminal network of sources and resistors can be replaced by a series combination of a single resistor R_{th} and voltage source V_{th}. This is represented by the example in Fig. 1.10. The sources can include both voltage and current sources (the current source is described below). A more general version of the theorem replaces the word *resistor* with *impedance*, a concept we will develop in Chapter 2.

The point of Thevenin's theorem is that when we connect a component to the terminals, it is much easier to analyze the circuit on the right than the circuit on the left. But there is no free lunch – we must first determine the values of V_{th} and R_{th}.

V_{th} is the voltage across the circuit terminals when nothing is connected to the terminals. This is clear from the equivalent circuit: if nothing is connected to the terminals, then no current flows in the circuit and there is no voltage drop across R_{th}. The voltage across the terminals is thus the same as V_{th}. In practice, the voltage across the terminals must be calculated by analyzing the original circuit.

There are two methods for calculating R_{th}; you can use whichever is easiest. In the first method, you start by short circuiting all the voltage sources and open circuiting all the current sources in the original circuit. This means that you replace the voltage sources by a wire and disconnect the current sources. Now only resistors are left in the circuit. These are then combined into one resistor using the resistor equivalent circuit laws. This one resistor then gives the value of R_{th}. In the second method, we calculate the current that would flow in the circuit if we shorted (placed a wire across) the terminals. Call this the short circuit current I_{sc}. Then from the Thevenin equivalent circuit it is clear that $R_{th} = \frac{V_{th}}{I_{sc}}$.

There is also a similar result known as Norton's theorem. This theorem states that any two-terminal network of sources and resistors can be replaced by a parallel

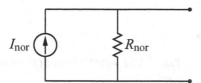

Figure 1.11 Equivalent circuit of Norton's theorem.

combination of a single resistor R_{nor} and current source I_{nor}. This equivalent circuit is shown in Fig. 1.11.

The current source is usually less familiar that the voltage source, but the two can be viewed as complements of one another. An ideal voltage source will maintain a constant voltage across it and will provide whatever current is required by the rest of the circuit. Similarly, an ideal current source will maintain a constant current through it while the voltage across it will be set by the rest of the circuit.

Returning now to the equivalent circuit, let's determine R_{nor} and I_{nor}. If we short the terminals, it is clear from the Norton equivalent circuit that all of I_{nor} will pass through the shorting wire. Thus $I_{nor} = I_{sc}$. We have seen previously that the voltage across the terminals when nothing is connected is equal to V_{th}. From the Norton equivalent circuit we then see that $V_{th} = I_{nor}R_{nor}$, so

$$R_{nor} = \frac{V_{th}}{I_{nor}} = \frac{V_{th}}{I_{sc}} = R_{th}. \tag{1.19}$$

1.2.2 Applications for resistors

Resistors are probably the most common circuit element and can be used in a variety of simple circuits. Here are a few examples.

1. Current limiting. Many electronic devices come with operating specifications. For example, the ubiquitous LED (light emitting diode) typically operates with a voltage drop of 1.7 V and a current of 20 mA. Suppose you have a 9 V battery and wish to light the LED. How can you operate the 1.7 V LED with a 9 V battery? By the discriminating use of a resistor! Consider the circuit in Fig. 1.12. KVL gives $V_0 - IR - V_{LED} = 0$, where V_{LED} is the voltage across the LED. We know that $V_0 = 9$ V, $V_{LED} = 1.7$ V, and we want $I = 20$ mA for proper operation. Solving for R gives

$$R = \frac{V_0 - V_{LED}}{I} = \frac{9 - 1.7 \text{ V}}{20 \times 10^{-3} \text{ A}} = 365 \ \Omega. \tag{1.20}$$

This is an example of using a resistor as a current limiter. Without it, the LED would burn out immediately.

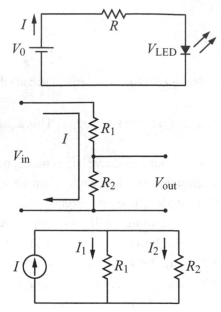

Figure 1.12 Application of a resistor as a current limiter.

Figure 1.13 The ubiquitous voltage divider.

Figure 1.14 The current divider.

2. Voltage divider. Another very common resistor circuit is shown in Fig. 1.13. Some voltage V_{in} is applied to the input and the circuit provides a lower voltage at the output. The analysis is simple. KVL gives $V_{in} = I(R_1 + R_2)$ and Ohm's Law gives $V_{out} = IR_2$. Solving for I from the first equation and substituting in the second gives

$$V_{out} = IR_2 = \left(\frac{V_{in}}{R_1 + R_2} \right) R_2 = V_{in} \left(\frac{R_2}{R_1 + R_2} \right) \qquad (1.21)$$

where this last form emphasizes that $V_{out} < V_{in}$ since $\frac{R_2}{R_1+R_2} < 1$. This equation is used so frequently it is worth memorizing.

3. The current divider circuit is shown in Fig. 1.14. A current source is applied to two resistors in parallel and we would like to obtain an expression that tells us how the current is divided between the two. By KCL, $I = I_1 + I_2$. Since the two resistors are in parallel, the voltage across them must be the same. Hence, $I_1 R_1 = I_2 R_2$. Solving this latter equation for I_2 and plugging into the first gives

$$I = I_1 + I_1 \left(\frac{R_1}{R_2} \right) = I_1 \left(\frac{R_1 + R_2}{R_2} \right) \qquad (1.22)$$

or

$$I_1 = \left(\frac{R_2}{R_1 + R_2} \right) I. \qquad (1.23)$$

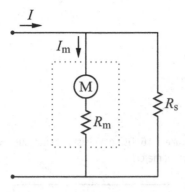

Figure 1.15 Using a resistor to extend the range of a current meter.

4. Multi-range analog voltmeter/ammeter. In electronics, one frequently has the need to measure voltage and current. The instrument of choice for many experimentalists is the multimeter, which can measure voltage, current, and resistance. The analog version of the multimeter uses a simple meter as a display. If you tear one of these multimeters apart, you find that the meter is a current measuring device that gives a full scale deflection of the needle for a given, small current, typically 50 μA. This is fine if you want to measure currents from zero to 50 μA, but what if you have a larger current to measure, or want to measure a voltage instead?

   Both of these can be accomplished by judicious use of resistors. The circuit in Fig. 1.15 shows a meter in parallel with a so-called shunt resistor R_s. The physical meter (within the dotted lines) is represented by an ideal current measuring meter in series with a resistor R_m. When a current I is applied to the terminals, part goes through the meter and part through the shunt. The circuit is simply a current divider, so we have (cf. Eq. (1.22))

$$I = I_m \left(1 + \frac{R_m}{R_s} \right). \tag{1.24}$$

A full scale deflection of the meter will always occur when $I_m = 50$ μA, and R_m is also set at the construction of the meter, but by adjusting the shunt resistance R_s we can make this full scale deflection occur for any input current I we choose.

   Another simple addition will allow us to use our meter to measure voltage. Placing a resistor R_s in series with the meter gives the configuration in Fig. 1.16. It is convenient here to define the voltage $V_m = I_m R_m$ that will produce a full scale deflection when applied across the physical meter. This circuit is then seen to be a voltage divider. Inverting Eq. (1.21) then gives

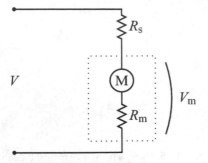

Figure 1.16 Using a resistor to measure voltage with a current meter.

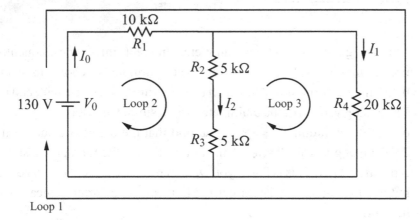

Figure 1.17 The standard method of solving circuit problems.

$$V = V_{\mathrm{m}} \left(1 + \frac{R_{\mathrm{s}}}{R_{\mathrm{m}}} \right) \tag{1.25}$$

so by varying R_{s} we can make the full scale deflection of the meter correspond to any input voltage.

1.2.3 Techniques for solving circuit problems

We list here three methods for solving circuit problems, and illustrate the use of these techniques on the same problem that we solved previously using equivalent circuit laws for resistors. Our goal is to solve for the current through resistor R_4 in Fig. 1.17.

The standard method This method involves assigning currents to each branch of the circuit and then applying KVL and KCL. In Fig. 1.17 we have assigned currents I_0, I_1, and I_2. In this case, the application of KCL gives a single equation

$$I_0 = I_1 + I_2 \tag{1.26}$$

but in circuits with more than three branches KCL gives additional relations. Next we use KVL around the loops indicated in the figure. For Loop 1 we get

$$V_0 - I_0 R_1 - I_1 R_4 = 0 \tag{1.27}$$

while Loop 2 gives

$$V_0 - I_0 R_1 - I_2(R_2 + R_3) = 0 \tag{1.28}$$

and finally

$$-I_1 R_4 + I_2(R_2 + R_3) = 0 \tag{1.29}$$

for Loop 3. We thus have four equations relating the three unknown currents I_0, I_1, and I_2 and need to solve for I_1. In practice we need only three independent equations to solve for the currents, but we have given all four here to illustrate the method. Solving Eq. (1.26) for I_2 (one of the currents we are not interested in) and plugging into Eq. (1.29) gives

$$-I_1 R_4 + (I_0 - I_1)(R_2 + R_3) = 0 \tag{1.30}$$

and solving Eq. (1.27) for I_0 gives

$$I_0 = \frac{V_0 - I_1 R_4}{R_1}. \tag{1.31}$$

Plugging Eq. (1.31) into Eq. (1.30) and solving for I_1 gives, after some algebra,

$$I_1 = \frac{V_0(R_2 + R_3)}{R_1 R_4 + (R_1 + R_4)(R_2 + R_3)}. \tag{1.32}$$

Plugging the circuit values into this equation gives $I_1 = 2.6$ mA, our former answer.

The mesh loop method Our second method for solving circuit problems is the mesh loop method. In this method, currents are assigned to the circuit loops rather than the actual physical branches of the circuit. This is shown in Fig. 1.18 where we assign current I_1 to the outer loop and I_2 to the inner loop.

We then move around these loops, applying KVL, but including contributions from both loop currents. The outer loop then gives

$$V_0 - (I_1 + I_2)R_1 - I_1 R_4 = 0 \tag{1.33}$$

while the inner loop gives

$$V_0 - (I_1 + I_2)R_1 - I_2(R_2 + R_3) = 0. \tag{1.34}$$

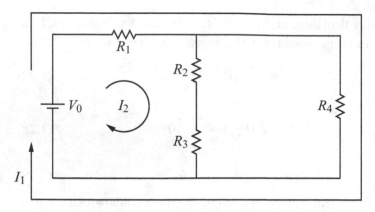

Figure 1.18 The mesh loop method of solving circuit problems.

Note that the resulting set of equations is simpler in this method: two equations in two unknowns I_1 and I_2. For this reason the mesh loop method is usually preferable for more complicated circuits. Furthermore, our equations can be rearranged into the conventional form of a system of linear algebraic equations. Thus Eq. (1.33) becomes

$$(R_1 + R_4)I_1 + R_1 I_2 = V_0 \tag{1.35}$$

while Eq. (1.34) gives

$$R_1 I_1 + (R_1 + R_2 + R_3)I_2 = V_0. \tag{1.36}$$

Students of linear algebra may wish to solve these using Cramer's Method of Determinants or with the built-in capabilities of many hand-held calculators (see Appendix B). The usual brute force method also works: solving Eq. (1.36) for I_2, plugging this into Eq. (1.35), and solving for I_1 produces (again, after some algebra),

$$I_1 = \frac{V_0(R_2 + R_3)}{(R_1 + R_4)(R_2 + R_3) + R_1 R_4}, \tag{1.37}$$

the same expression obtained with the standard method.

Thevenin's theorem Finally, we solve this problem by using Thevenin's theorem. We form the required two terminal network by removing R_4 and taking the two terminals at the points where R_4 was attached. This is shown in Fig. 1.19.

The remaining circuit should look familiar – if we combine R_2 and R_3 it is the previously considered voltage divider. Thus

$$V_{\text{th}} = V_0 \left(\frac{R_2 + R_3}{R_1 + R_2 + R_3} \right) = 65 \text{ V}. \tag{1.38}$$

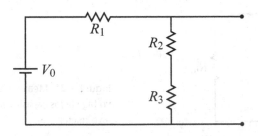

Figure 1.19 First step in solving the problem using Thevenin's theorem.

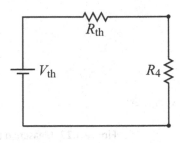

Figure 1.20 Last step: re-attach R_4 to the Thevenin equivalent circuit.

Shorting out the battery leaves R_1 in parallel with $R_2 + R_3$ so

$$R_{th} = \frac{R_1(R_2 + R_3)}{R_1 + R_2 + R_3} = 5\,k\Omega. \tag{1.39}$$

Reattaching R_4 then gives the simple circuit of Fig. 1.20 with the current through R_4 given by

$$I_1 = \frac{V_{th}}{R_{th} + R_4} = \frac{65\,V}{25\,k\Omega} = 2.6\,mA \tag{1.40}$$

as before.

1.2.4 Input resistance

A common measurement in the electronics lab is the voltage across a component. An important fact to keep in mind when making such measurements is that *the measuring instrument becomes part of the circuit*. The act of measuring thus inevitably changes the thing we are trying to measure because we are adding circuitry to the original circuit. To help us cope with this problem, test instrument manufacturers specify a quantity called the *input resistance* R_{in} (or, as we will see later, the *input impedance*). The effect of attaching the instrument is the same as attaching a resistor with value R_{in}. To see how this helps, suppose we are measuring the voltage across some resistor R_0 in a complicated circuit, as depicted in Fig. 1.21.

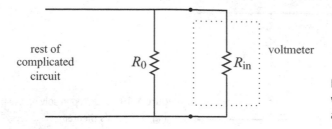

Figure 1.21 Measuring the voltage across resistor R_0 with a voltmeter.

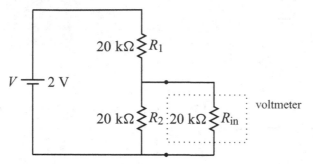

Figure 1.22 Measuring the output of a voltage divider with a voltmeter.

If we know the input resistance of our measuring device we see that the effect of making the measurement is to replace the original resistor R_0 with the parallel combination of R_0 with R_{in}

$$R_0 \rightarrow R_0 \parallel R_{in} = \frac{R_0 R_{in}}{R_0 + R_{in}} \tag{1.41}$$

where $R_0 \parallel R_{in}$ is shorthand for the parallel combination. From this, one can see that the circuit-altering effect of attaching the measuring instrument is mitigated by making the input resistance as high as possible, because

$$\frac{R_0 R_{in}}{R_0 + R_{in}} \rightarrow R_0 \tag{1.42}$$

as $R_{in} \rightarrow \infty$.

As an example of what happens when R_{in} is not large, consider the circuit in Fig. 1.22. Ignoring the meter for a moment we see that the original circuit is a voltage divider, and application of Eq. (1.21) gives $V_{out} = 1$ V. But the effect of the meter's input resistance is to change R_2 to $R_2 \parallel R_{in} = 10$ kΩ. Using this in Eq. (1.21) gives $V_{out} = \frac{2}{3}$ V, and this is what the meter will indicate. So, unless we are aware of the effect of input resistance, we run the danger of making a false measurement. On the other hand, if we are aware of this effect, we can analyze the effect and determine the true value of our voltage when the meter is unattached.

Figure 1.23 Impedance specification for a typical analog meter.

How does one determine the value of the input resistance for a given instrument? Here are some common ways.

1. Look in the instrument manual under *input resistance* or *input impedance*. The value should be in ohms.
2. For analog voltmeters, look for a specification with units of ohms per volt (Ω/V). This is usually printed on the face of the meter itself, as shown in Fig. 1.23. To get R_{in}, multiply this number by the full scale voltage selected. For example, suppose your meter is specified as 20000 Ω/V and you have selected the 2.5 V full scale setting. The input resistance is then $20000 \times 2.5 = 50$ kΩ.
3. You may have to analyze the instrument circuitry itself. The relevant question is: when a voltage is applied to the input of the instrument, how much current flows into the instrument? Then, by Ohm's Law, the input resistance is just the ratio of this voltage and current.

1.3 AC signals

So far our examples have used constant voltage sources such as batteries. Constant voltages and currents are described as DC quantities in electronics. On the other

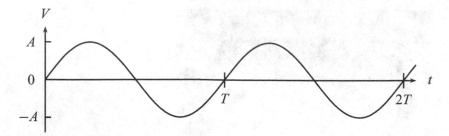

Figure 1.24 A sine wave.

hand, voltages and currents that vary in time are called AC quantities. For future reference, we list here some of the most common AC signals.

1. Sinusoidal signals. This is probably the most fundamental signal in electronics since, as we will see later, any signal can be constructed from sinusoidal signals. A typical sinusoidal voltage is shown in Fig. 1.24.

   Sinusoidal voltages can be written

$$V = A \sin(2\pi ft + \phi) = A \sin(\omega t + \phi) \tag{1.43}$$

where A is the amplitude, f is the frequency in cycles/second or hertz (abbreviated Hz), ϕ is the phase, and ω is the angular frequency (in radians/second). The repetition time t_{rep} is also called the period T of the signal, and this is related to the frequency of the signal by $T = \frac{1}{f}$.

   There are several ways to specify the amplitude of a sinusoidal signal that are in common use. These include the following.

   (a) The peak amplitude A or A_p.

   (b) The peak-to-peak amplitude $A_{pp} = 2A$.

   (c) The rms amplitude $A_{rms} = A/\sqrt{2}$. This is useful for power calculations involving sinusoidal waves. For example, suppose we want the power dissipated in a resistor given the sinusoidally varying voltage across it. We cannot simply use Eq. (1.8) since our voltage is varying in time (what V would we use?). Instead, we calculate the time average of the power over one period:

$$P = \frac{1}{T} \int_0^T \frac{V^2}{R} dt = \frac{1}{TR} \int_0^T A^2 \sin^2(\omega t + \phi) dt = \frac{A^2}{2R} = \frac{A_{rms}^2}{R}. \tag{1.44}$$

This last form shows that we *can* use Eq. (1.8) to calculate the power as long as we use the rms amplitude of the sinusoidal signal in the formula. The same argument applies to Eq. (1.7) for sinusoidal currents.

(d) Decibels (abbreviated dB) are used to compare the amplitude of two signals, say A_1 and A_2:

$$\mathrm{dB} = 20 \log_{10} \frac{A_2}{A_1} = 10 \log_{10} \left(\frac{A_2}{A_1}\right)^2 = 10 \log_{10} \frac{P_2}{P_1} \qquad (1.45)$$

where this last expression uses the power level of the two signals. So, for example, if $A_2 = 2A_1$, then we get $20 \log 2 \approx 6$, so we say A_2 is 6 dB higher than A_1. Various related schemes specify the decibel level relative to a fixed standard. So dBV is the dB relative to a 1 V_{rms} signal and dBm is the dB relative to a 0.78 V_{rms} signal. For the curious, this latter voltage standard is 1 mW into a 600 Ω resistor.

Some other typical waveforms of electronics are shown in Figs. 1.25 through 1.30.

2. Square wave. Specified by an amplitude and a frequency (or period).

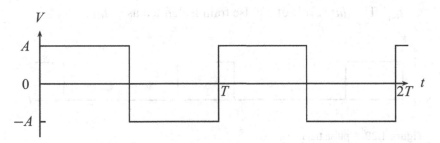

Figure 1.25 The square wave.

3. Sawtooth wave. Specified by an amplitude and a frequency (or period).

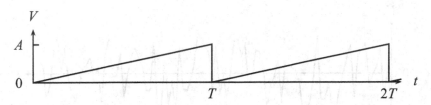

Figure 1.26 The sawtooth wave.

4. Triangle wave. Specified by an amplitude and a frequency (or period).

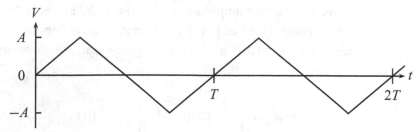

Figure 1.27 The triangle wave.

5. Ramp. Specified by an amplitude and a ramp time.

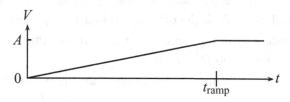

Figure 1.28 A ramp signal.

6. Pulse train. Specified by an amplitude, a pulse width τ, and a repetition time t_{rep}. The *duty cycle* of a pulse train is defined as τ/t_{rep}.

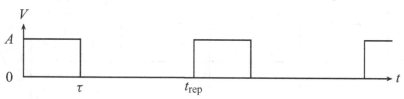

Figure 1.29 A pulse train.

7. Noise. These are random signals of thermal origin or simply unwanted signals coupled into the circuit. Noise is usually described by its frequency content, but that is a more advanced topic.

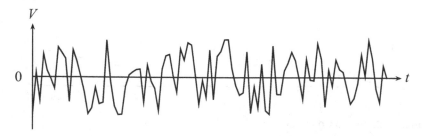

Figure 1.30 Noise.

EXERCISES

1. What is the resistance of a nichrome wire 1 mm in diameter and 1 m in length?
2. What is the maximum allowable current through a $10\,k\Omega$, $10\,W$ resistor? Through a $10\,k\Omega$, $1/4\,W$ resistor?
3. (a) What power rating is needed for a $100\,\Omega$ resistor if $100\,V$ is to be applied to it? (b) For a $100\,k\Omega$ resistor?
4. Compute the current through R_3 of Fig. 1.31.

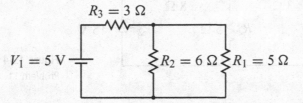

$R_3 = 3\,\Omega$

$V_1 = 5\,V$ $R_2 = 6\,\Omega$ $R_1 = 5\,\Omega$

Figure 1.31 Circuit for Problems 4 and 5.

5. Compute the current through R_1 and R_2 of Fig. 1.31.
6. The output of the voltage divider of Fig. 1.32 is to be measured with voltmeters with input resistances of $100\,\Omega$, $1\,k\Omega$, $50\,k\Omega$, and $1\,M\Omega$. What voltage will each indicate?

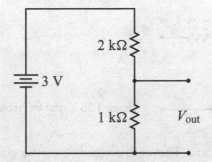

$2\,k\Omega$

$3\,V$

$1\,k\Omega$ V_{out}

Figure 1.32 Circuit for Problem 6.

7. A real battery can be modeled as an ideal voltage source in series with a resistor (the *internal resistance*). A voltmeter with input resistance of $1000\,\Omega$ measures the voltage of a worn-out $1.5\,V$ flashlight battery as $0.9\,V$. What is the internal resistance of the battery?
8. If the flashlight battery of the preceding problem had been measured with a voltmeter with input resistance of $10\,M\Omega$, what would the reading be?
9. What is the resistance across the terminals of Fig. 1.33?
10. Suppose that a $25\,V$ battery is connected to the terminals of Fig. 1.33. Find the current in the $10\,\Omega$ resistor.
11. Compute the current through R_2 and R_3 of Fig. 1.34.

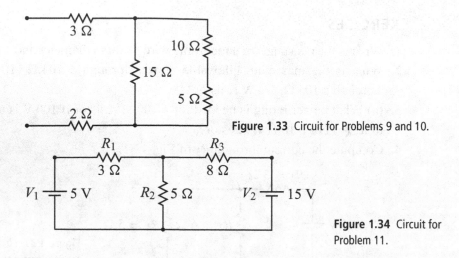

Figure 1.33 Circuit for Problems 9 and 10.

Figure 1.34 Circuit for Problem 11.

12. Find the Thevenin voltage and Thevenin resistance of the circuit shown in Fig. 1.35.

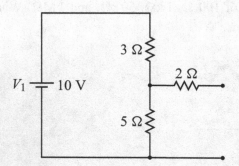

Figure 1.35 Circuit for Problem 12.

13. Find the Thevenin voltage and Thevenin resistance of the circuit shown in Fig. 1.36 with R_5 removed. The two terminals for this problem are the points where R_5 was connected.

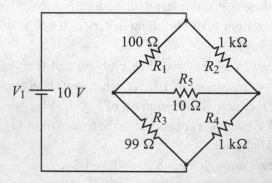

Figure 1.36 Circuit for Problems 13 and 14.

14. Using the result of the previous problem, find the current through R_5 of Fig. 1.36.

15. In the circuit of Fig. 1.37, compute the current in the 3 Ω resistor and find the value of V_2.

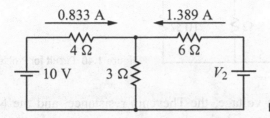

Figure 1.37 Circuit for Problem 15.

16. In the circuit of Fig. 1.38, find the value of V_3 such that the current in the 10 Ω resistor is zero.

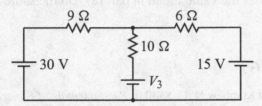

Figure 1.38 Circuit for Problem 16.

17. Compute all the currents labeled in the circuit of Fig. 1.39 assuming the following values: $V_1 = 5\,\text{V}$, $V_2 = 10\,\text{V}$, $V_3 = 15\,\text{V}$, $R_1 = 2\,\Omega$, $R_2 = 4\,\Omega$, $R_3 = 6\,\Omega$, $R_4 = 7\,\Omega$, $R_5 = 5\,\Omega$, $R_6 = 3\,\Omega$. Suggestion: use the mesh loop method.

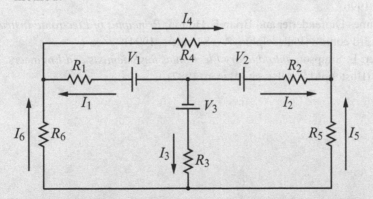

Figure 1.39 Circuit for Problem 17.

18. (a) Compute the current through the 10 Ω resistor in the circuit of Fig. 1.40. Do not use Thevenin's or Norton's theorems for this computation. (b) Now

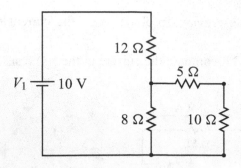

Figure 1.40 Circuit for Problem 18.

find the Thevenin voltage, the Thevenin resistance, and the Norton current when the 10 Ω resistor is removed. The two terminals for this problem are the points where the 10 Ω resistor was connected. (c) Show that, if the 10 Ω resistor is connected to the Thevenin equivalent circuit, the current through the 10 Ω resistor matches the value found in part (a). Do the same for the Norton equivalent circuit.

FURTHER READING

Charles K. Alexander and Matthew N. O. Sadiku, *Fundamentals of Electric Circuits*, 2nd edition (New York: McGraw-Hill, 2004).

L. W. Anderson and W. W. Beeman, *Electric Circuits and Modern Electronics* (New York: Holt, Rinehart, and Winston, 1973).

Dennis Barnaal, *Analog Electronics for Scientific Application* (Prospect Heights, IL: Waveland Press, 1989).

James J. Brophy, *Basic Electronics for Scientists*, 5th edition (New York: McGraw-Hill, 1990).

A. James Diefenderfer and Brian E. Holton, *Principles of Electronic Instrumentation*, 3rd edition (Philadelphia, PA: Saunders, 1994).

Robert E. Simpson, *Introductory Electronics for Scientists and Engineers*, 2nd edition (Boston, MA: Allyn and Bacon, 1987).

2 AC circuits

2.1 Introduction

Currents and voltages that vary in time are called *AC* quantities. When analyzing circuits where the current and voltage change in time, the treatment of resistors is unchanged: they still obey Ohm's Law. In this chapter we introduce two other basic circuit components, the capacitor and the inductor. The treatment of these components depends on the details of how things are changing in time, and this will require the development of some new analysis techniques.

2.2 Capacitors

Another basic circuit component is the capacitor. A capacitor is formed by any pair of conductors, but the usual form is two parallel plates. For this case, the capacitance C is given by

$$C = \epsilon \frac{A}{d} \tag{2.1}$$

where A is the area of a plate, d is the distance between plates, and ϵ is the dielectric constant of the material between the plates. Note that, like the resistance, the capacitance depends only on the physical characteristics of the device. The unit of capacitance is coulombs per volt or *farads*, abbreviated F. Typical capacitor values are in a range such that µF or pF are convenient units. When purchasing a capacitor, you must specify its *voltage rating* in addition to its capacitance value. This rating tells you the maximum voltage you can apply across the capacitor before there is electrical breakdown through the dielectric material.

So what does a capacitor do? One answer is that it is a charge storage device. When a voltage V is applied to a capacitor, a charge of magnitude Q will be stored on each plate. Q is given by

$$Q = CV. \tag{2.2}$$

Figure 2.1 Equivalent circuit for capacitors in series.

In electronics, we are usually concerned with currents (the *flow* of charge) rather than charge. If we take the time derivative of Eq. (2.2) and note that, by definition, $I = \frac{dQ}{dt}$ we get

$$I = C\frac{dV}{dt}. \tag{2.3}$$

Viewed from this perspective, C is the constant relating a time-varying voltage across the capacitor to the AC current through the capacitor.

2.2.1 Equivalent circuit laws for capacitors

As with resistors, capacitors in series and parallel can be combined to form simpler equivalent circuits.

2.2.1.1 Series capacitors

Consider, for example, three capacitors in series as shown in Fig. 2.1. We wish to combine the capacitors to form the equivalent circuit on the right.

Let Q_1 be the charge on capacitor C_1 and so on. Applying KVL and Eq. (2.2) we obtain

$$V - \frac{Q_1}{C_1} - \frac{Q_2}{C_2} - \frac{Q_3}{C_3} = 0. \tag{2.4}$$

By charge conservation, the charge on each capacitor is the same, so $Q_1 = Q_2 = Q_3 \equiv Q$ and

$$V = Q\left(\frac{1}{C_1} + \frac{1}{C_2} + \frac{1}{C_3}\right). \tag{2.5}$$

Comparing this with Eq. (2.2), we see that the equivalent capacitance C_{eq} will be given by

$$\frac{1}{C_{eq}} = \frac{1}{C_1} + \frac{1}{C_2} + \frac{1}{C_3} \tag{2.6}$$

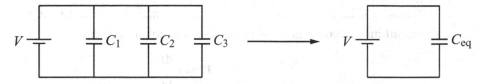

Figure 2.2 Equivalent circuit for capacitors in parallel.

or, generalizing this to any number of capacitors in series,

$$\frac{1}{C_{eq}} = \sum_i \frac{1}{C_i}. \tag{2.7}$$

2.2.1.2 Parallel capacitors

Now consider three capacitors in parallel as shown in Fig. 2.2. Again, let Q_1 be the charge on capacitor C_1 and so on. Because the capacitors are connected in parallel, the voltage across them must be the same

$$V = \frac{Q_1}{C_1} = \frac{Q_2}{C_2} = \frac{Q_3}{C_3}. \tag{2.8}$$

If we add the three charges and apply Eq. (2.8) to each term, we get

$$Q_1 + Q_2 + Q_3 = V(C_1 + C_2 + C_3). \tag{2.9}$$

If we are to form an equivalent capacitor the battery must supply the same amount of charge in both cases. Thus $Q_{eq} = Q_1 + Q_2 + Q_3$ and, from Eq. (2.9), $Q_{eq} = V(C_1 + C_2 + C_3)$. Comparing this with Eq. (2.2), we see that

$$C_{eq} = C_1 + C_2 + C_3 \tag{2.10}$$

or, generalizing this to any number of capacitors in parallel,

$$C_{eq} = \sum_i C_i. \tag{2.11}$$

Note that capacitors combine opposite to the way resistors combine: *series* resistors add up directly while *parallel* capacitors add up directly.

2.3 Inductors

We learn in introductory physics that currents produce magnetic fields (Ampère's Law) and that time-varying magnetic fields can induce a voltage in a circuit

(Faraday's Law). Putting these together means that time-varying currents in a circuit induce voltages. This is expressed in equation form by

$$V = L\frac{\mathrm{d}I}{\mathrm{d}t} \tag{2.12}$$

where the constant L is called the *self-inductance* or simply the inductance. While any circuit loop has inductance, we usually ignore this (like we ignore the small resistance of connecting wires) and, if inductance is required in a circuit, add discrete inductors made of coils of wire. For a long coil (i.e., a solenoid), this inductance is given by

$$L = \frac{\mu N^2 \pi R^2}{l} \tag{2.13}$$

where μ is the permeability of the material on which the coil is wound, N is the number of turns in the coil, R is the radius of the coil, and l is the length of the coil. The unit of inductance is volts times seconds per amp or *henries* (abbreviated H).

The derivation of the equivalent circuit laws for inductors in series and parallel is similar to that for resistors, and we leave the details to the reader. The result for inductors in series is

$$L_{\mathrm{eq}} = \sum_i L_i \tag{2.14}$$

and for inductors in parallel

$$\frac{1}{L_{\mathrm{eq}}} = \sum_i \frac{1}{L_i}. \tag{2.15}$$

Note that inductors combine the same way that resistors do.

2.4 RC circuits

Now we turn to our first circuit that combines components – in this case a resistor and capacitor in series. Consider the circuit in Fig. 2.3.

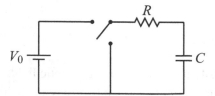

Figure 2.3 A switched RC circuit.

The resistor-capacitor combination is connected to a switch which can be positioned to connect to the battery V_0 or to a short. We can analyze both cases at once: applying KVL we obtain

$$V = IR + \frac{Q}{C} \tag{2.16}$$

where V is equal to V_0 when the switch connects to the battery and is equal to zero when the switch is down. We next take the derivative of Eq. (2.16) in order to remove Q in favor of the current I. Since V is constant for either switch position, we obtain for both cases

$$0 = R\frac{dI}{dt} + \frac{I}{C}. \tag{2.17}$$

We thus have a first order differential equation for the current I that must be solved to complete the analysis of the circuit. Rearranging Eq. (2.17), we obtain

$$\frac{dI}{I} = -\frac{dt}{RC}. \tag{2.18}$$

Integrating both sides gives

$$\ln I = -\frac{t}{RC} + K \tag{2.19}$$

where K is the constant of integration. Finally, we exponentiate both sides

$$I = \exp\left(-\frac{t}{RC} + K\right) = I_0 \exp\left(-\frac{t}{RC}\right). \tag{2.20}$$

In this last step we have introduced a new constant I_0 in place of the constant $\exp(K)$. Equation (2.20) is the general solution for the current as a function of time. We will see how to determine the constant I_0 in a moment.

The voltage across the resistor is just IR. Let's also determine the voltage across the capacitor, V_c. KVL gives (for either switch position) $V_c = V - IR$. Employing Eq. (2.20) we obtain

$$V_c = V - I_0 R \exp\left(-\frac{t}{RC}\right). \tag{2.21}$$

For future reference, we note that, since V and I_0R are constants, we can write a general solution for V_c as

$$V_c = V_1 \exp\left(-\frac{t}{RC}\right) + V_2 \tag{2.22}$$

where V_1 and V_2 are constants.

2.4.1 Charging

To further determine the behavior of the circuit, we need to specify the conditions when the switch is thrown (i.e., the *initial conditions*). Suppose we assemble the circuit with an uncharged capacitor so that V_c is initially zero. We define $t = 0$ at the instant we throw the switch to connect the battery. Thus, at $t = 0$, $V_c = 0$ and $V = V_0$. Using this information in Eq. (2.21), we obtain:

$$0 = V_0 - I_0 R \tag{2.23}$$

which then gives us the unknown constant $I_0 = \frac{V_0}{R}$. Using this information in Eqs. (2.20) and (2.21) gives us our specific solutions:

$$I = \frac{V_0}{R} \exp\left(-\frac{t}{RC}\right) \tag{2.24}$$

and

$$V_c = V_0 \left[1 - \exp\left(-\frac{t}{RC}\right)\right]. \tag{2.25}$$

These solutions are plotted in Fig. 2.4. It is worth noting some of the key features. The current starts at its maximum value V_0/R and then falls toward zero. The capacitor starts (as we specified) with zero voltage and approaches the battery voltage V_0 as it charges up. Each solution has an exponential term with a decay that depends on the ratio t/RC. The product RC, which has units of time, is called the *time constant* of the decay. It determines how long it takes for the circuit to approach its final state. When $t = RC$, the current has dropped to about 37% of its initial value and the capacitor has reached roughly 63% of its final value. This notion of time constant makes physical sense, too: a larger capacitor will take longer to fill with charge; a larger resistor will limit the flow of charge, thus increasing the time it takes to charge up the capacitor.

2.4.2 Discharging

Suppose that we have waited long enough that the capacitor has become fully charged. We now throw the switch to the down position. Resetting our clock to $t = 0$, our initial conditions are now $V_c = V_0$ and $V = 0$. Using this information in Eq. (2.21) gives:

$$V_0 = 0 - I_0 R \rightarrow I_0 = -\frac{V_0}{R} \tag{2.26}$$

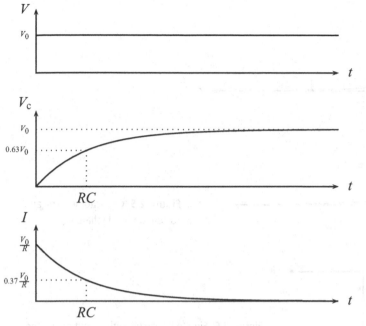

Figure 2.4 Capacitor voltage and current for RC charging.

so our specific solutions become

$$I = -\frac{V_0}{R} \exp\left(-\frac{t}{RC}\right) \tag{2.27}$$

and

$$V_c = V_0 \exp\left(-\frac{t}{RC}\right). \tag{2.28}$$

These solutions are plotted in Fig. 2.5. As the capacitor discharges, the capacitor voltage decays exponentially with a time constant RC and approaches zero (the voltage level it is now attached to). The current is negative because it flows in the opposite direction during discharge; it also decays exponentially with the same time constant.

2.4.3 Response to a square wave

We can use the insight we have obtained from our study of the switched RC circuit to sketch out the response of an RC circuit to a square wave drive. The square wave is a little different than our switch problem in that the voltage switches between V_0 and $-V_0$ rather than V_0 and zero, so in the square wave case the discharge will approach $-V_0$.

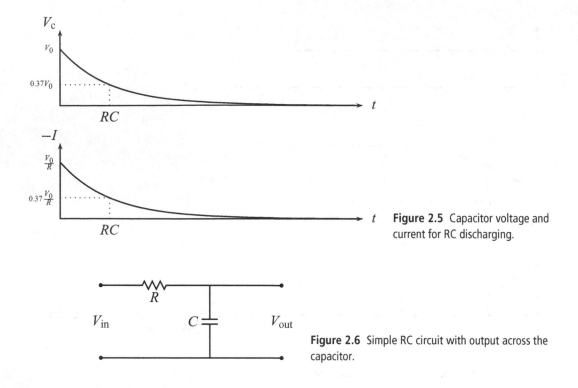

Figure 2.5 Capacitor voltage and current for RC discharging.

Figure 2.6 Simple RC circuit with output across the capacitor.

Let's first examine the case where the output of the circuit is the voltage across the capacitor. We can represent this case with the circuit diagram of Fig. 2.6 where V_{in} is a square wave of amplitude V_0 and period T. The shape of the output voltage will depend on the relative size of RC and $T/2$. Some representative cases are shown in Fig. 2.7.

If $RC \ll T/2$, the capacitor has plenty of time to charge up fully while the square wave voltage stays constant. We thus see the same shape waveform as we saw in the switch problem. When the input voltage switches from V_0 to $-V_0$, the capacitor discharges (or charges to the opposite polarity) and approaches the new input voltage. Note that for this case, the output voltage looks like a square wave with rounded leading edges.

If $RC \approx T/2$, the capacitor initially charges toward V_0, but only gets part of the way there before the input switches to $-V_0$. It now tries to charge to this input voltage, but, again, does not have the time to get there. Now the output voltage is quite different from a square wave. The maximum and minimum voltages are not shown for this case since they depend on the exact relationship between RC and $T/2$, but we can say that the maximum voltage is less than V_0 and the minimum voltage is greater than $-V_0$.

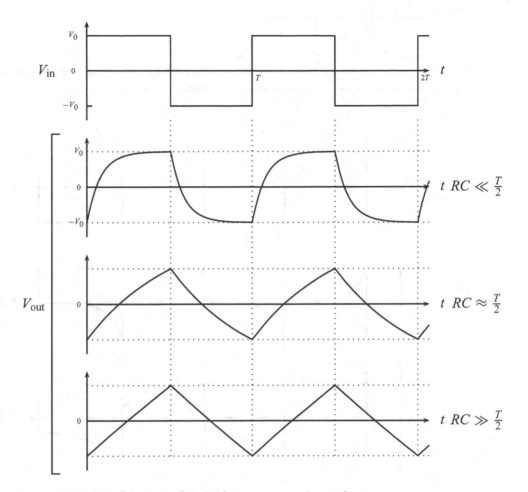

Figure 2.7 Output of the circuit of Fig. 2.6 for a square-wave input voltage.

If $RC \gg T/2$, the capacitor has even less time to charge and discharge before the input voltage switches. In this case, the waveform is a series of rising and falling lines forming a triangle wave. This reflects the fact that the first term in the expansion of $[1 - \exp(-t/RC)]$ for small t/RC is linear in time.

2.4.4 Voltage across the resistor

Having put so much work into our analysis of the RC circuit, let's squeeze out another result. Suppose we apply our square wave to the rearranged circuit of Fig. 2.8. Now our output voltage is the voltage across the resistor. This may seem like new territory, but it really is not since $V_{out} = IR$ and we already solved the

AC circuits

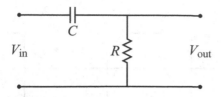

Figure 2.8 Simple RC circuit with output across the resistor.

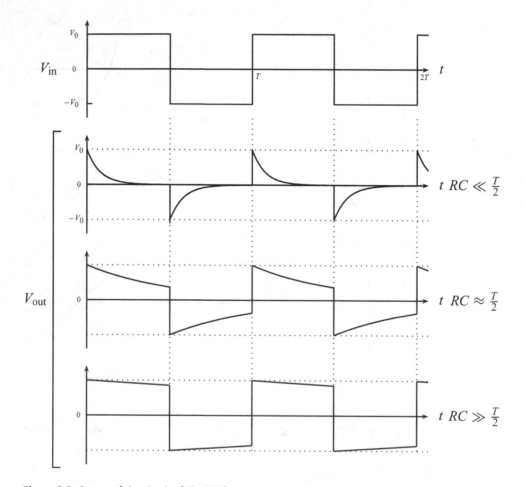

Figure 2.9 Output of the circuit of Fig. 2.8 for a square-wave input voltage.

problem for I. Again, the shape of the output voltage depends on the relative size of RC and $T/2$ as shown in Fig. 2.9.

When $RC \ll T/2$, we have the case that is most like our switching problem. The current (and thus the output voltage) starts at its maximum value and decays exponentially toward zero. Since $RC \ll T/2$, there is plenty of time to decay all the way to zero. When the input voltage switches polarity, the current goes negative and again decays toward zero. The output voltage resembles a series of positive and negative spikes.

As the RC time constant becomes larger with respect to $T/2$, there is not enough time for the current to decay all the way to zero before the input switches, so the output voltage waveforms take on the shapes shown in the figure. When $RC \gg T/2$, the spiky behavior is gone and the output looks like a distorted square wave.

2.5 Response to a sine wave

We now consider the response of an RC circuit to a sine wave drive. While it may appear that we are just making a minor change to the input waveform, the change is actually much more profound. For the switching case (or the square wave), the input voltage was, at each instant of time, constant. This allowed us to solve the differential equation (Eq. (2.17)) resulting from the circuit analysis, producing simple functions of time for I and V_c. This approach is called a *time domain analysis*. For more complicated input voltages, time domain analysis is not always possible because we cannot solve the resulting differential equation. In these cases, it is sometimes useful to analyze the circuit in terms of its sine wave response, which we will call a *frequency domain analysis*.

The relevant RC circuit is shown in Fig. 2.10, where now the voltage source is a sine wave input V_{in}. As usual, KVL gives

$$V_{in} = \frac{Q}{C} + IR \tag{2.29}$$

and taking the time derivative of this and rearranging yields

$$R\frac{dI}{dt} + \frac{I}{C} = \frac{dV_{in}}{dt}. \tag{2.30}$$

Note that, unlike the switching problem, the derivative of the input voltage is not zero. To proceed, we now specify the input voltage as $V_{in} = V_p \sin \omega t$ and assume

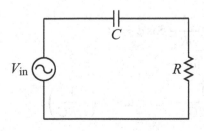

Figure 2.10 Simple RC circuit driven by a sinusoidal voltage.

the current has the form $I = I_p \sin(\omega t + \phi)$, where I_p and ϕ are constants to be determined. Plugging into Eq. (2.30) results in

$$R\omega I_p \cos(\omega t + \phi) + \frac{I_p}{C} \sin(\omega t + \phi) = \omega V_p \cos \omega t. \qquad (2.31)$$

Note that our approach has allowed us to turn a differential equation (2.30) into an algebraic equation (2.31). We now proceed to solve for I_p and ϕ.

In order to isolate the unknown constants ϕ and I_p, we employ the following trig identities:

$$\sin(\omega t + \phi) = \sin \omega t \cos \phi + \cos \omega t \sin \phi \qquad (2.32)$$

and

$$\cos(\omega t + \phi) = \cos \omega t \cos \phi - \sin \omega t \sin \phi. \qquad (2.33)$$

Dividing Eq. (2.31) by $R\omega I_p$ and applying these identities gives

$$(\cos \omega t \cos \phi - \sin \omega t \sin \phi) + \frac{1}{\omega RC}(\sin \omega t \cos \phi + \cos \omega t \sin \phi) = \frac{V_p}{I_p R} \cos \omega t.$$
$$(2.34)$$

Rearranging, we get

$$\left(\cos \phi + \frac{1}{\omega RC} \sin \phi - \frac{V_p}{I_p R}\right) \cos \omega t + \left(-\sin \phi + \frac{1}{\omega RC} \cos \phi\right) \sin \omega t = 0.$$
$$(2.35)$$

In order to proceed in our quest for I_p and ϕ, we make the following argument: Eq. (2.35) is valid for all times t, and thus must be valid for any particular time we choose. If we choose $t = 0$, then $\sin \omega t = 0$ and $\cos \omega t = 1$, and Eq. (2.35) reduces to

$$\cos \phi + \frac{1}{\omega RC} \sin \phi - \frac{V_p}{I_p R} = 0. \qquad (2.36)$$

Alternatively, if we choose t such that $\omega t = \frac{\pi}{2}$, then $\sin \omega t = 1$ and $\cos \omega t = 0$ and we obtain

$$-\sin \phi + \frac{1}{\omega RC} \cos \phi = 0. \qquad (2.37)$$

This last equation can now be solved for ϕ:

$$\frac{\sin \phi}{\cos \phi} = \tan \phi = \frac{1}{\omega RC} \rightarrow \phi = \tan^{-1}\left(\frac{1}{\omega RC}\right). \qquad (2.38)$$

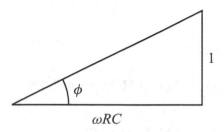

Figure 2.11 Right triangle satisfying Eq. (2.38).

Equation (2.36) requires $\sin\phi$ and $\cos\phi$ in order to solve for I_p. If we have values for ω, R, and C then, of course, we can obtain a number for ϕ from Eq. (2.38) and simply plug-in to get $\sin\phi$ and $\cos\phi$. But we prefer to obtain general algebraic results, so we use the following trick. A right triangle satisfying Eq. (2.38) is shown in Fig. 2.11. It thus follows that $\sin\phi$ and $\cos\phi$ are given by

$$\sin\phi = \frac{1}{\sqrt{1 + (\omega RC)^2}} \tag{2.39}$$

and

$$\cos\phi = \frac{\omega RC}{\sqrt{1 + (\omega RC)^2}}. \tag{2.40}$$

Using these expressions in Eq. (2.36),

$$\frac{\omega RC}{\sqrt{1 + (\omega RC)^2}} + \frac{1}{\omega RC}\frac{1}{\sqrt{1 + (\omega RC)^2}} = \frac{V_p}{I_p R} \tag{2.41}$$

which gives, after some algebra,

$$I_p = \frac{\omega C}{\sqrt{1 + (\omega RC)^2}} V_p. \tag{2.42}$$

Recalling the form we assumed for the current at the beginning, our final solution is

$$I = \frac{\omega C V_p}{\sqrt{1 + (\omega RC)^2}} \sin(\omega t + \phi), \tag{2.43}$$

where ϕ is given by

$$\phi = \tan^{-1}\left(\frac{1}{\omega RC}\right). \tag{2.44}$$

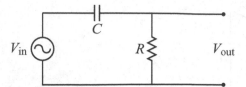

Figure 2.12 A high-pass RC filter.

2.5.1 RC positive phase shifter and high-pass filter

We can now apply our result to two common circuits. The first, shown in Fig. 2.12, takes the output voltage V_{out} across the resistor.

Since we have already solved for the current, getting this output voltage is easy:

$$V_{\text{out}} = IR = \frac{\omega RCV_{\text{p}}}{\sqrt{1 + (\omega RC)^2}} \sin(\omega t + \phi). \tag{2.45}$$

Recalling that our input voltage has the assumed form $V_{\text{in}} = V_{\text{p}} \sin \omega t$, we see that the output voltage has changed in two ways. (1) It has shifted in phase. Since $\omega RC > 0$, Eq. (2.44) tells us that $0 < \phi < \frac{\pi}{2}$. Thus our circuit is a *positive phase shifter*. (2) The amplitude has changed. It is useful here to ignore the time variation and phase shift and simply compare the magnitude of the input signal $|V_{\text{in}}|$ with the magnitude of the output $|V_{\text{out}}|$:

$$\frac{|V_{\text{out}}|}{|V_{\text{in}}|} = \frac{\omega RC}{\sqrt{1 + (\omega RC)^2}}. \tag{2.46}$$

It is clear from the form of Eq. (2.46) that the relative output amplitude depends only on the product ωRC. We examine the easily calculable extreme limits to get some idea of the behavior. When $\omega RC \to 0$, $|V_{\text{out}}|/|V_{\text{in}}| \to 0$, and when $\omega RC \to \infty$, $|V_{\text{out}}|/|V_{\text{in}}| \to 1$. Plotting Eq. (2.46) gives us the full picture (see Fig. 2.13).

The plot shows that lower frequencies (giving lower ωRC) are attenuated, that is, the output amplitude of such signals is much smaller than the input amplitude. Higher frequencies, on the other hand, are relatively unattenuated and pass through the circuit with little change in their amplitude. This behavior is characteristic of a *high-pass filter*. The *breakpoint frequency* or *half-power frequency*, defined by $\omega RC = 1$ and shown in the plot, gives the point below which attenuation starts to be significant.

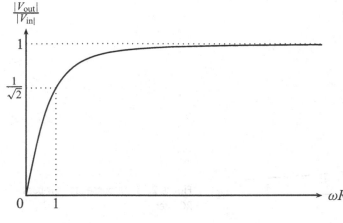

Figure 2.13 Response of a high-pass RC filter.

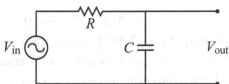

Figure 2.14 A low-pass RC filter.

2.5.2 RC negative phase shifter and low-pass filter

Another common use of our RC circuit can be obtained by taking the output voltage across the capacitor as shown in Fig. 2.14. In this case $V_{out} = Q/C$, but $Q = \int I \, dt$, so

$$Q = I_p \int \sin(\omega t + \phi) \, dt = -\frac{I_p}{\omega} \cos(\omega t + \phi) \qquad (2.47)$$

where we have taken the integration constant to be zero since we assume there is no DC charge on the capacitor. Using our former result for I_p (Eq. (2.42)):

$$V_{out} = \frac{-V_p}{\sqrt{1 + (\omega RC)^2}} \cos(\omega t + \phi) = \frac{V_p}{\sqrt{1 + (\omega RC)^2}} \sin\left(\omega t + \phi - \frac{\pi}{2}\right) \qquad (2.48)$$

where in the last step we have used the identity $\cos A = -\sin\left(A - \frac{\pi}{2}\right)$ so as to better compare the result with the form of the input voltage $V_{in} = V_p \sin \omega t$. As before, $0 < \phi < \frac{\pi}{2}$, so $-\frac{\pi}{2} < \phi - \frac{\pi}{2} < 0$. This circuit is thus a *negative phase shifter*. Looking at the relative magnitudes:

$$\frac{|V_{out}|}{|V_{in}|} = \frac{1}{\sqrt{1 + (\omega RC)^2}}. \qquad (2.49)$$

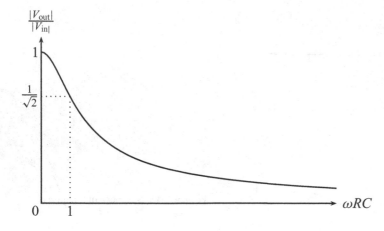

Figure 2.15 Response of a low-pass RC filter.

Here the extreme cases give $|V_{out}|/|V_{in}| \to 1$ as $\omega RC \to 0$, , and $|V_{out}|/|V_{in}| \to 0$ as $\omega RC \to \infty$. The full plot of Eq. (2.49) is given in Fig. 2.15.

In this case the plot shows that *lower* frequencies are *un*attenuated while the higher frequencies are suppressed. This circuit is therefore a *low-pass filter*. The breakpoint frequency is defined, as before, by $\omega RC = 1$.

2.5.3 The RC circuit as differentiator and integrator

Now that we have some insight into the behavior of the RC circuit we can perform one more analysis that gives yet another usage for the circuit. Consider first the high-pass circuit given above. KVL gives

$$V_{in} - \frac{Q}{C} - V_{out} = 0 \tag{2.50}$$

or $Q = C(V_{in} - V_{out})$. Taking the time derivative of this latter equation gives

$$I = C\frac{d}{dt}(V_{in} - V_{out}) = \frac{V_{out}}{R} \tag{2.51}$$

where in the last equality we have used the fact that $V_{out} = IR$. Now from Eq. (2.46) we see that, for small ωRC, $|V_{out}| \ll |V_{in}|$, so we can ignore the second term of the derivative in Eq. (2.51). Using this fact and rearranging we get

$$V_{out} \approx RC\frac{dV_{in}}{dt} \tag{2.52}$$

where the approximation holds when ωRC is small. Thus, under these conditions, our high-pass filter functions as a *differentiator circuit*, giving as output a voltage that is proportional to the derivative of the input voltage.

A similar analysis can be applied to the low-pass circuit. Combining KVL, $I = dQ/dt$, and $Q = CV_{out}$ we find

$$I = \frac{V_{in} - V_{out}}{R} = \frac{dQ}{dt} = C\frac{dV_{out}}{dt}. \tag{2.53}$$

This time Eq. (2.49) tells us that, for large ωRC, $|V_{out}| \ll |V_{in}|$, so we can ignore V_{out} in the first equality of Eq. (2.53). Hence $V_{in} \approx RC(dV_{out}/dt)$ or

$$V_{out} \approx \frac{1}{RC} \int V_{in}\, dt \tag{2.54}$$

where the approximation holds when ωRC is large. Under these conditions, our low-pass filter functions as an *integrator circuit*, giving as output a voltage that is proportional to the integral of the input voltage.

Lastly, we note that these conclusions are consistent with the response of the RC circuit to a square-wave input voltage that we studied earlier. The output taken across the capacitor (which we now recognize as the integrator configuration) gave a triangle wave when $RC \gg T/2$ (cf. Fig. 2.7). The triangle is indeed the integral of the square wave since the integral of a constant is a linear function rising or falling according to the sign of the constant. We get the triangle wave only when $RC \gg T/2$ because this is when the approximation of Eq. (2.54) holds (since $\omega = 2\pi/T$). Similarly, when the output was taken across the resistor (now recognized as the differentiator circuit) we obtained a series of positive and negative spikes when $RC \ll T/2$ (cf. Fig. 2.9). These spikes approximate the delta-function derivatives of the square wave and the condition on the relative size of RC and $T/2$ is just the differentiator condition of small ωRC.

2.6 Using complex numbers in electronics

2.6.1 Introduction

We have seen that one way to solve our circuit differential equations is to assume currents of the form $I = I_p \sin(\omega t + \phi)$. This transforms the differential equation into an algebraic equation which we can then solve. This, however, involves considerable work and requires that we know certain trig identities. We can solve these problems more easily by employing complex numbers. This approach also has the advantage of producing a broader conceptual understanding of resistors, capacitors, and inductors.

2.6.2 The basics of complex numbers

You are probably already familiar with several different types of numbers: integers, rational numbers, and real numbers. Each of these types has rules that apply to the manipulation of the member numbers: addition, multiplication, exponentiation, etc. A complex number is simply another type of number system with its own set of rules for manipulation.

In general, a complex number $\hat{z}$ can be written as

$$\hat{z} = a + jb, \tag{2.55}$$

where a and b are real numbers and $j \equiv \sqrt{-1}$. Note that we signify a complex number by the "hat" symbol (^), and that we use j for the square root of -1 rather than the more common i to avoid confusion with the symbol for current. Complex numbers can be manipulated using the same algebraic rules as for real numbers, except now we have, in addition, a way of representing the square root of a negative number, e.g., $\sqrt{-3.2} = j\sqrt{3.2}$.

It is sometimes convenient to represent a complex number as a point on the complex plane, with the vertical axis being the imaginary part of $\hat{z}$ (written $\text{Im}(\hat{z})$), and the horizontal axis being the real part of $\hat{z}$ (written $\text{Re}(\hat{z})$). Such a representation is shown in Fig. 2.16. The point can also be represented using the length $|\hat{z}|$ of the line from the origin to the point and the angle θ this line makes with the positive horizontal axis:

$$\hat{z} = |\hat{z}|(\cos\theta + j\sin\theta). \tag{2.56}$$

$|\hat{z}|$ and θ are also referred to as the *magnitude* and *phase* of $\hat{z}$. The two representations are related by the equations

$$|\hat{z}| = \sqrt{a^2 + b^2} \tag{2.57}$$

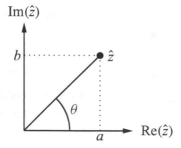

Figure 2.16 A complex number as represented by a point in the complex plane.

and

$$\theta = \tan^{-1}\left(\frac{b}{a}\right). \tag{2.58}$$

Equation (2.56) can be cast in a more useful form by employing the Taylor expansions for sine, cosine, and the exponential:

$$\cos\theta + j\sin\theta = \left(1 - \frac{\theta^2}{2!} + \frac{\theta^4}{4!} - \cdots\right) + j\left(\theta - \frac{\theta^3}{3!} + \frac{\theta^5}{5!} - \cdots\right)$$
$$= 1 + j\theta + \frac{(j\theta)^2}{2!} + \frac{(j\theta)^3}{3!} + \frac{(j\theta)^4}{4!} + \cdots$$
$$= e^{j\theta}. \tag{2.59}$$

Using this in Eq. (2.56), we see that any complex number can also be written in the so-called complex exponential form:

$$\hat{z} = |\hat{z}|e^{j\theta}, \tag{2.60}$$

where $|\hat{z}|$ and θ are related to the the real and imaginary parts of $\hat{z}$ by Eqs. (2.57) and (2.58).

Finally, we note a common definition. If we have a complex number of the form $\hat{z} = a + jb$, then the *complex conjugate* $\hat{z}^*$ of this number is defined as $\hat{z}^* = a - jb$. Multiplying any complex number by its complex conjugate gives a real number equal to the magnitude of the number squared:

$$\hat{z}\hat{z}^* = (a + jb)(a - jb) = a^2 - j^2b^2 = a^2 + b^2 = |\hat{z}|^2. \tag{2.61}$$

2.6.3 The series RC circuit

We now use complex numbers to re-solve the problem of a series RC circuit (Fig. 2.17). We use the same V_{in} as before (except now we use cosine rather than sine), but now note that this can be written as the real part of a complex number:

$$V_{in} = V_p \cos\omega t = \text{Re}(V_p e^{j\omega t}). \tag{2.62}$$

Similarly, we write the current as:

$$I = I_p \cos(\omega t + \phi) = \text{Re}(I_p e^{j(\omega t+\phi)}) = \text{Re}(\hat{I}_p e^{j\omega t}) \tag{2.63}$$

where in the last equation we have written $I_p e^{j\phi}$ as the complex current amplitude $\hat{I}_p$.

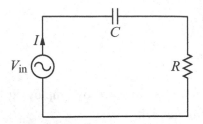

Figure 2.17 Series RC circuit.

We now use these complex voltages and currents to solve the circuit equation. This, of course, is a cheat. The voltage we apply to the circuit is a real number, not a complex one. We deal with this objection by agreeing to take the real part of our answer when we finish the solution. As before, applying Kirchoff's Laws to the circuit gives

$$R\frac{dI}{dt} + \frac{I}{C} = \frac{dV_{in}}{dt}. \tag{2.64}$$

We now substitute in the complex voltage $V_p e^{j\omega t}$ and current $\hat{I}_p e^{j\omega t}$, and obtain

$$R\hat{I}_p j\omega e^{j\omega t} + \frac{1}{C}\hat{I}_p e^{j\omega t} = j\omega V_p e^{j\omega t}. \tag{2.65}$$

Canceling the $e^{j\omega t}$ and dividing by $j\omega$, we have

$$\left(R + \frac{1}{j\omega C}\right)\hat{I}_p = V_p \tag{2.66}$$

and solving for the current amplitude $\hat{I}_p$ gives

$$\hat{I}_p = \frac{V_p}{\left(R + \frac{1}{j\omega C}\right)} = \frac{\omega C V_p}{\omega RC - j} \tag{2.67}$$

where in the last step we have multiplied top and bottom by ωC and used the fact that $\frac{1}{j} = -j$. It remains to massage the answer into a nicer form. We use the fact, noted above, that any complex number can be written in complex exponential form with the magnitude and angle given by Eqs. (2.57) and (2.58), respectively. Applying this to the denominator of Eq. (2.67):

$$\omega RC - j = \sqrt{(\omega RC)^2 + (-1)^2}\, e^{j\theta} \tag{2.68}$$

where θ is given by

$$\tan \theta = \frac{-1}{\omega RC}. \tag{2.69}$$

The complex current amplitude (2.67) can then be written as

$$\hat{I}_p = \frac{\omega C V_p}{\sqrt{(\omega RC)^2 + 1}}\, e^{-j\theta}. \tag{2.70}$$

Finally, as agreed, we take the real part of our result:

$$I = \mathrm{Re}(\hat{I}_p e^{j\omega t}) = \frac{\omega C V_p}{\sqrt{1 + (\omega RC)^2}}\mathrm{Re}(e^{j(\omega t - \theta)}) = \frac{\omega C V_p}{\sqrt{1 + (\omega RC)^2}}\cos(\omega t - \theta) \tag{2.71}$$

which is the same as our former result.

2.6.4 Discussion and generalization

What have we gained by using complex numbers? At a minimum, we now have another method that allows us to solve the differential equations arising from the analysis of LRC circuits with sinusoidal drive voltages. To my taste, the complex exponential method is easier because it does not require the trig identities and algebraic tricks of our former method. The price, of course, is learning to use and manipulate complex numbers.

There is, however, an additional advantage and a new conceptual insight that comes from the use of complex numbers. Look at Eq. (2.66). It looks vaguely like Ohm's Law for resistors: something times current equals voltage. But this circuit involves a capacitor as well as a resistor, whereas Ohm's Law only applies to resistors. A similar result occurs when we analyze circuits that involve inductors. It turns out that it is possible to generalize Ohm's Law to include capacitors and inductors on equal footing with resistors.

To do this, we introduce the concept of *impedance*. Impedance is a generalization of resistance that applies to resistors, capacitors, and inductors alike. Generally, the impedance of a component or circuit is a complex number, and we use the symbol $\hat{Z}$ to denote it. The impedances of our three components are

- $\hat{Z}_{\text{resistor}} = R$
- $\hat{Z}_{\text{capacitor}} = \frac{1}{j\omega C}$
- $\hat{Z}_{\text{inductor}} = j\omega L.$

An impedance *impedes* or limits the flow of current and is thus a generalized resistance. Note that, for capacitors and inductors, the impedance depends on the frequency ω. When $\omega \to 0$, $Z_{\text{capacitor}} \to \infty$, which is consistent with our knowledge that no DC current flows through a capacitor. On the other hand, when

$\omega \rightarrow \infty$, $Z_{\text{inductor}} \rightarrow \infty$, so an inductor cannot pass high frequency currents. Finally, some vocabulary. The real part of a complex impedance is called the *resistive impedance* or simply the *resistance*, while the imaginary part is called the *reactive impedance* or simply the *reactance*. The reactance is often given the symbol χ.

Circuit analysis is also simplified by this unified approach to resistors, capacitors, and inductors. Since impedances act like generalized resistors, the rules for series and parallel combinations of impedances are the same as those for resistors:

$$\hat{Z}_{\text{series}} = \sum_i \hat{Z}_i \qquad (2.72)$$

and

$$\frac{1}{\hat{Z}_{\text{parallel}}} = \sum_i \frac{1}{\hat{Z}_i}. \qquad (2.73)$$

The final simplification comes from the fact that we no longer have to deal with differential equations. Circuit analysis is reduced to applications of a generalized complex Ohm's Law:

$$\hat{V} = \hat{I}\hat{Z}. \qquad (2.74)$$

2.6.5 Applications

We now apply the complex Ohm's Law to several circuits. For cases where we wish to find the current produced by a drive voltage, we can reduce the technique to a recipe:

- recall the complex Ohm's Law $\hat{V} = \hat{I}\hat{Z}$
- find the total circuit impedance $\hat{Z}_{\text{tot}}$
- calculate $\hat{I} = \frac{\hat{V}}{\hat{Z}_{\text{tot}}}$
- massage the resulting complex numbers into the form $a + jb$
- convert this number into complex exponential form $|\hat{z}|e^{j\theta}$
- plug in the appropriate $\hat{V}$
- take the real part of the resulting complex current $\hat{I}$.

2.6.5.1 Series RC circuit
First let's apply the recipe to the previously studied RC circuit. The total impedance of this series combination of a resistor and capacitor is simply $\hat{Z}_{\text{tot}} = R + \frac{1}{j\omega C}$.

Hence we have

$$\hat{I} = \frac{\hat{V}}{R + \frac{1}{j\omega C}}. \tag{2.75}$$

This equation is very similar to Eq. (2.67), and the "massaging" is identical to what we did there. The result is

$$\hat{I} = \frac{\omega C \hat{V}}{\sqrt{(\omega RC)^2 + 1}} \, e^{-j\theta}, \tag{2.76}$$

where, as before, θ is given by

$$\theta = \tan^{-1}\left(\frac{-1}{\omega RC}\right). \tag{2.77}$$

In most cases, we have a drive voltage of the form $V_p \cos(\omega t)$ or $V_p \sin(\omega t)$. In either case, we can plug into Eq. (2.76) a complex voltage $V_p e^{j\omega t}$ with the proviso that, at the end, we will take the real part of the answer for the cosine drive or the imaginary part of the answer for the sine drive. Thus

$$\hat{I} = \frac{\omega C V_p}{\sqrt{(\omega RC)^2 + 1}} \, e^{j(\omega t - \theta)} \tag{2.78}$$

and, for a cosine drive, we obtain our former result:

$$I = \text{Re}(\hat{I}) = \frac{\omega C V_p}{\sqrt{(\omega RC)^2 + 1}} \cos(\omega t - \theta). \tag{2.79}$$

2.6.5.2 Series LR circuit

Now let's apply the technique to a circuit we have not studied before, the series LR circuit shown in Fig. 2.18. The total impedance of this series combination of a resistor and inductor is $\hat{Z}_{\text{tot}} = R + j\omega L$. Hence,

$$\hat{I} = \frac{\hat{V}}{R + j\omega L} = \frac{\hat{V}}{\sqrt{R^2 + (\omega L)^2} \, e^{j\theta}} \tag{2.80}$$

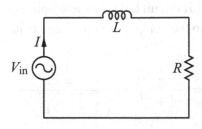

Figure 2.18 Series LR circuit.

where

$$\theta = \tan^{-1}\left(\frac{\omega L}{R}\right). \tag{2.81}$$

As before, we use $\hat{V} = V_{\mathrm{p}}e^{\mathrm{j}\omega t}$ and obtain

$$\hat{I} = \frac{V_{\mathrm{p}}}{\sqrt{R^2 + (\omega L)^2}}\,e^{\mathrm{j}(\omega t - \theta)}. \tag{2.82}$$

If the quantity sought is the current, we simply take the real part of Eq. (2.82) and we are done. Often, however, the circuit's function is to produce an output voltage which is obtained by taking the voltage across one of the components. This can be obtained by multiplying the current times the impedance of the component. For example, suppose we want the voltage across the resisitor as the output. Then,

$$\hat{V}_{\mathrm{out}} = \hat{I}\hat{Z}_{\mathrm{resistor}} = \frac{V_{\mathrm{p}}R}{\sqrt{R^2 + (\omega L)^2}}\,e^{\mathrm{j}(\omega t - \theta)} \tag{2.83}$$

and, taking the real part,

$$V_{\mathrm{out}} = \frac{V_{\mathrm{p}}R}{\sqrt{R^2 + (\omega L)^2)}}\cos(\omega t - \theta). \tag{2.84}$$

Since θ as given by Eq. (2.81) will be positive, this circuit acts as a *negative phase shifter*. It also acts as a frequency filter, as can be seen by looking at the relative magnitudes of the input and output voltages

$$\frac{|V_{\mathrm{out}}|}{|V_{\mathrm{in}}|} = \frac{\frac{V_{\mathrm{p}}R}{\sqrt{R^2 + (\omega L)^2}}}{V_{\mathrm{p}}} = \frac{1}{\sqrt{1 + \left(\frac{\omega L}{R}\right)^2}}. \tag{2.85}$$

We can see from this last form that the relative output amplitude has a maximum at $\omega = 0$ and falls off to zero as $\omega \to \infty$. The entire curve is shown in Fig. 2.19. Note that low frequency signals pass through unattenuated while higher frequency signals (those higher than R/L) are reduced in amplitude. Thus, when the output is taken across the resistor, the circuit thus acts as a *low-pass filter*.

Suppose now that we use the same LR circuit but take the output voltage across the inductor. This voltage, according to the complex Ohm's Law, is just $\hat{I}\hat{Z}_{\mathrm{inductor}}$. Thus, from Eq. (2.80) we have

$$\hat{V}_{\mathrm{out}} = \frac{\mathrm{j}\omega L\hat{V}}{R + \mathrm{j}\omega L} = \frac{\omega L\hat{V}}{-\mathrm{j}R + \omega L} = \frac{\frac{\omega L}{R}\hat{V}}{-\mathrm{j} + \frac{\omega L}{R}} = \frac{\frac{\omega L}{R}\hat{V}}{\sqrt{1 + \left(\frac{\omega L}{R}\right)^2}\,e^{\mathrm{j}\theta}} \tag{2.86}$$

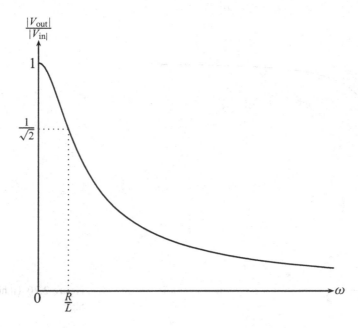

Figure 2.19 LR low-pass filter.

where now

$$\theta = \tan^{-1}\left(\frac{-1}{\frac{\omega L}{R}}\right) = \tan^{-1}\left(\frac{-R}{\omega L}\right). \tag{2.87}$$

Proceeding as before, we get

$$V_{\text{out}} = \frac{\frac{\omega L}{R} V_{\text{p}}}{\sqrt{1 + \left(\frac{\omega L}{R}\right)^2}} \cos(\omega t - \theta) \tag{2.88}$$

and

$$\frac{|V_{\text{out}}|}{|V_{\text{in}}|} = \frac{\frac{\omega L}{R}}{\sqrt{1 + \left(\frac{\omega L}{R}\right)^2}}. \tag{2.89}$$

Since θ is now negative (see Eq. (2.87)), the circuit acts as a *positive* phase shifter. As before, it also acts as a frequency filter, but now it is a *high*-pass filter, as can be seen by looking at the graph of Eq. (2.89) shown in Fig. 2.20.

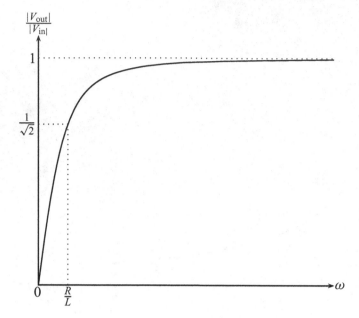

Figure 2.20 LR high-pass filter.

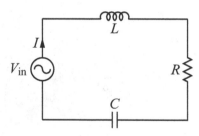

Figure 2.21 Series LRC circuit.

2.6.5.3 Series LRC circuit

As a final example, consider the series LRC circuit shown in Fig. 2.21. For this circuit, the total impedance is $\hat{Z}_{tot} = R + j\omega L + \frac{1}{j\omega C}$. Hence,

$$\hat{I} = \frac{\hat{V}}{R + j\omega L + \frac{1}{j\omega C}} = \frac{\hat{V}}{R + j\left(\omega L - \frac{1}{\omega C}\right)} = \frac{\hat{V}}{\sqrt{R^2 + \left(\omega L - \frac{1}{\omega C}\right)^2}\, e^{j\theta}} \qquad (2.90)$$

where

$$\theta = \tan^{-1}\left(\frac{\omega L - \frac{1}{\omega C}}{R}\right). \qquad (2.91)$$

Again using $\hat{V} = V_p e^{j\omega t}$, we obtain

$$\hat{I} = \frac{V_p}{\sqrt{R^2 + \left(\omega L - \frac{1}{\omega C}\right)^2}} e^{j(\omega t - \theta)} \qquad (2.92)$$

and taking the real part of $\hat{I}$ yields

$$I = \frac{V_p}{\sqrt{R^2 + \left(\omega L - \frac{1}{\omega C}\right)^2}} \cos(\omega t - \theta). \qquad (2.93)$$

The frequency dependence of I is of particular interest. Note that as $\omega \to 0$, $\frac{1}{\omega C} \to \infty$, and $|I| \to 0$. Also, as $\omega \to \infty$, $\omega L \to \infty$, and $|I| \to 0$. So for very low and very high frequencies we get no current. Somewhere in between there must be a maximum. We can find it by setting the derivative of the amplitude with respect to ω equal to zero:

$$0 = \frac{d}{d\omega} \frac{1}{\sqrt{R^2 + \left(\omega L - \frac{1}{\omega C}\right)^2}}$$

$$= -\frac{1}{2}\left[R^2 + \left(\omega L - \frac{1}{\omega C}\right)^2\right]^{-3/2} 2\left(\omega L - \frac{1}{\omega C}\right)\left(L + \frac{1}{\omega^2 C}\right) \qquad (2.94)$$

which has solution $\omega L = \frac{1}{\omega C}$ or

$$\omega = \frac{1}{\sqrt{LC}} \equiv \omega_0. \qquad (2.95)$$

The entire curve is shown in Fig. 2.22. Systems exhibiting a large response at a particular drive frequency are called *resonant systems*, and the frequency at which the response peaks is called the *resonant frequency*. Thus, for the series LRC circuit, the resonant frequency is ω_0. The current amplitude at that point is V_p/R, the value it would have if the inductor and capacitor were removed from the circuit. Apparently, at the resonant frequency, the effect of the inductive and capacitive impedance cancels. The width of the curve $\Delta\omega$ at a current $\frac{1}{\sqrt{2}}$ down from the peak can be shown to be approximately R/L, where the approximation is good for $R/L \ll \omega_0$. Thus, the smaller R is, the narrower the curve and the higher the peak current. If we take the voltage across the resistor as our output, $V_{out} = IR$ and the circuit functions as a *band-pass filter*, only allowing frequencies near ω_0 to pass through. Such circuits are routinely used to tune-in a selected communication channel while suppressing the neighboring transmission frequencies.

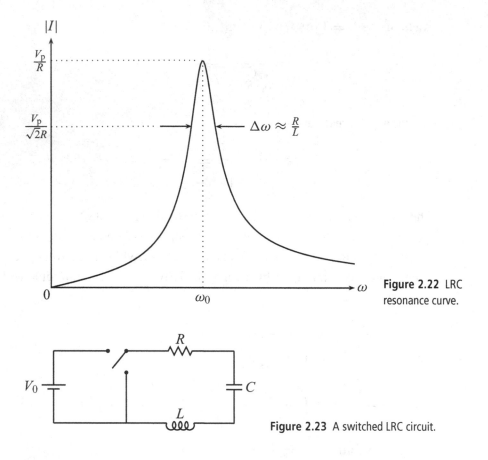

Figure 2.22 LRC resonance curve.

Figure 2.23 A switched LRC circuit.

2.7 Using the complex exponential method for a switching problem

While our method of complex exponentials was developed for circuits with a sinusoidal drive voltage, it can also be applied to switching problems. As an example, consider the circuit in Fig. 2.23.

Applying KVL, we obtain

$$IR + \frac{Q}{C} + L\frac{\mathrm{d}I}{\mathrm{d}t} = V \qquad (2.96)$$

where $V = V_0$ when the switch is up and $V = 0$ when the switch is down. Let's suppose that we are interested in the voltage across the capacitor, $V_c = Q/C$. It will be useful, therefore, to solve for the charge Q directly. Using $I = \mathrm{d}Q/\mathrm{d}t$, Eq. (2.96) becomes

$$L\frac{d^2Q}{dt^2} + R\frac{dQ}{dt} + \frac{Q}{C} = V. \tag{2.97}$$

Equation (2.97) is mathematically classified as a second order, linear, inhomogeneous differential equation. The general solution to such an equation is the sum of (1) the general solution to the homogeneous equation (i.e., Eq. (2.97) with the right hand side set to zero) and (2) any particular solution to Eq. (2.97). For the particular solution, we simply take the constant solution $Q = CV$. That leaves the homogeneous equation to solve. Using our complex exponential technique, we plug in $\hat{Q} = Q_p \exp(j\omega t)$. This gives

$$\left(-\omega^2 L + j\omega R + \frac{1}{C}\right) Q_p \exp(j\omega t) = 0. \tag{2.98}$$

Canceling $\hat{Q}$ and rearranging, we obtain a quadratic equation for ω:

$$\omega^2 - j\gamma\omega - \omega_0^2 = 0 \tag{2.99}$$

where we have defined $\gamma \equiv R/L$ and $\omega_0^2 \equiv 1/LC$. Solving for ω:

$$\omega = \frac{j\gamma \pm \sqrt{-\gamma^2 + 4\omega_0^2}}{2} = j\frac{\gamma}{2} \pm \sqrt{\omega_0^2 - \frac{\gamma^2}{4}}. \tag{2.100}$$

We thus obtain two values for ω. For the general solution we must use both.

2.7.1 Underdamped case

To go further with the solution we must specify the relative magnitudes of ω_0 and γ. We first consider the case where $\omega_0^2 > \frac{\gamma^2}{4}$, the so-called *underdamped case*. Since the square root in Eq. (2.100) is then a real number, our two values for ω are $\omega = j\frac{\gamma}{2} \pm \omega_1$, where we have defined

$$\omega_1 \equiv \sqrt{\omega_0^2 - \frac{\gamma^2}{4}}. \tag{2.101}$$

Using both values to form our solution gives

$$\hat{Q} = Q_1 e^{-\frac{\gamma}{2}t} e^{j\omega_1 t} + Q_2 e^{-\frac{\gamma}{2}t} e^{-j\omega_1 t} \tag{2.102}$$

where Q_1 and Q_2 are constants. Taking the real part:

$$Q = (Q_1 + Q_2)e^{-\frac{\gamma}{2}t} \cos\omega_1 t. \tag{2.103}$$

To this we must add our particular solution $Q = CV$ to obtain the general solution to Eq. (2.97). Since we are interested in the voltage across the capacitor V_c, we divide the result by C and obtain

$$V_c = V_3 e^{-\frac{\gamma}{2}t} \cos \omega_1 t + V \qquad (2.104)$$

where we have combined constants to form the new constant $V_3 \equiv (Q_1 + Q_2)/C$.

To complete the solution we must specify V_3 by applying the initial conditions. Assuming the capacitor is initially uncharged ($V_c = 0$ at $t = 0$) when the switch is placed in the up position ($V = V_0$), Eq. (2.104) yields $V_3 = -V_0$ and the solution becomes

$$V_c = V_0 \left(1 - e^{-\frac{\gamma}{2}t} \cos \omega_1 t\right). \qquad (2.105)$$

The solution has an oscillating part ($\cos \omega_1 t$), but the amplitude of this oscillation decays in time according to $\exp\left(-\frac{\gamma}{2}t\right)$. This behavior is known as *ringing* since it is reminiscent of a ringing bell sound. As $t \to \infty$, the solution approaches the constant V_0 and the capacitor becomes fully charged. This behavior is shown in Fig. 2.24.

Imagine now that, after the capacitor has become fully charged, we throw the switch to the down position so that $V_c = V_0$ at $t = 0$ and $V = 0$. Applying these conditions to Eq. (2.104) shows that $V_3 = V_0$, so the solution becomes

$$V_c = V_0 e^{-\frac{\gamma}{2}t} \cos \omega_1 t. \qquad (2.106)$$

Again, we have an oscillation at frequency ω_1 that dies off on a time scale set by γ. After a long enough time, the capacitor becomes fully discharged and $V_c = 0$. Combining this with the former result then gives the full behavior of the switched underdamped circuit as shown in Fig. 2.25.

This underdamped behavior is seen in other physical systems as well. For example, a lightly damped harmonic oscillator that has its equilibrium position suddenly changed will oscillate around that new equilibrium position until the oscillations are damped out.

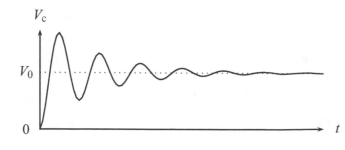

Figure 2.24 Ringing of an LRC circuit.

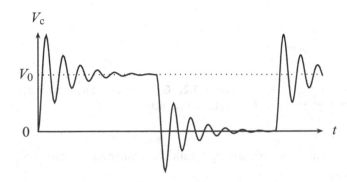

Figure 2.25 Underdamped response of a switched LRC circuit.

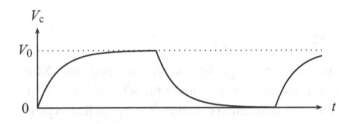

Figure 2.26 Overdamped response of a switched LRC circuit.

2.7.2 Overdamped case

Next, let us consider the case where $\omega_0^2 < \frac{\gamma^2}{4}$, the so-called *overdamped case*. Since the square root in Eq. (2.100) is now an imaginary number, our two solutions become $\omega = j\left(\frac{\gamma}{2} \pm \beta\right)$, where we have defined $\beta \equiv \sqrt{\gamma^2/4 - \omega_0^2}$. Again using both solutions to form our solution gives

$$\hat{Q} = Q_1 e^{-\frac{\gamma}{2}t} e^{-\beta t} + Q_2 e^{-\frac{\gamma}{2}t} e^{\beta t} \qquad (2.107)$$

where Q_1 and Q_2 are constants. Note that there is no danger of the second term diverging since $\beta < \frac{\gamma}{2}$ by construction. Proceeding as before we obtain

$$V_c = \frac{V_0}{2}\left(2 - e^{-\left(\frac{\gamma}{2}+\beta\right)t} - e^{-\left(\frac{\gamma}{2}-\beta\right)t}\right) \qquad (2.108)$$

for the charging portion of the solution and

$$V_c = \frac{V_0}{2}\left(e^{-\left(\frac{\gamma}{2}+\beta\right)t} + e^{-\left(\frac{\gamma}{2}-\beta\right)t}\right) \qquad (2.109)$$

for the discharging portion.

A plot of these functions is shown in Fig. 2.26. There are no oscillations in this case. Rather, there is a slow, monotonic approach to the final value. This is characteristic of a system where the damping (in this case produced by the circuit

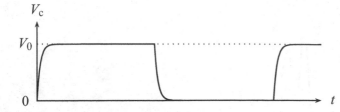

Figure 2.27 Critically damped response of a switched LRC circuit.

resistance) is large. The mechanical system analog would be a harmonic oscillator in a viscous fluid like molasses.

2.7.3 Critically damped case

Finally, let's look at what happens if $\omega_0^2 = \frac{\gamma^2}{4}$, the *critically damped case*. Now the square root in Eq. (2.100) is zero, and our two values for ω become identical. In such cases the methods of ordinary differential equations can be applied to yield a second independent solution. The resulting solution will have the form

$$V_c = (A + Bt)e^{-\frac{\gamma}{2}t} + V \tag{2.110}$$

with A and B constants. This critically damped case, shown in Fig. 2.27, produces the most rapid change to the new equilibrium value of V_c.

2.8 Fourier analysis

We have seen that the complex exponential method gives us a powerful method for solving problems involving combinations of resistors, capacitors, and inductors when the drive voltage is sinusoidal. But what about other types of drive voltage (e.g., triangle, sawtooth, etc.)? As we will now show, our analysis is applicable to *any* periodic signal as a consequence of a remarkable theorem named for Joseph Fourier.

The theorem is easily stated. Let $f(t)$ be any real, periodic function with period T such that $f(t) = f(t + T)$ for any t. Then there exist complex constants $\hat{c}_n$ such that

$$f(t) = \sum_{n=-\infty}^{\infty} \hat{c}_n e^{j\omega_n t} \tag{2.111}$$

where

$$\omega_n = \frac{2\pi n}{T} \equiv n\omega_1 \qquad (2.112)$$

and the constants are given by

$$\hat{c}_n = \frac{1}{T} \int_{t'}^{t'+T} f(t)e^{-j\omega_n t}dt \qquad (2.113)$$

for all n and any t'. Note that, as a consequence of Eq. (2.113), it is also true that $c_{-n} = c_n^*$.

Some powerful insights result from consideration of this theorem. Up to now, we have solved our circuit problems by assuming a voltage drive of the form $V_p \exp(j\omega t)$, so we know how to handle this case. But Eq. (2.111) says that *any* periodic function can be written as the sum of terms of this form, so we, in principle, can deal with the periodic function by dealing with each term on the right hand side of Eq. (2.111). To see how this works, let's return to the example of the series LRC circuit. Suppose now that we have a drive of the form $\hat{V}_n = \hat{c}_n \exp(j\omega_n t)$. Then, as before, the complex Ohm's Law gives

$$\hat{c}_n e^{j\omega t} = \hat{I}_n \hat{Z}_{\mathrm{tot}} = \hat{I}\left(R + j\omega_n L + \frac{1}{j\omega_n C}\right). \qquad (2.114)$$

We solve this for $\hat{I}_n$ as before and obtain

$$\hat{I}_n = \frac{\hat{c}_n}{\sqrt{R^2 + \left(\omega_n L - \frac{1}{\omega_n C}\right)^2}}e^{j(\omega_n t - \phi_n)} \qquad (2.115)$$

where

$$\phi_n = \tan^{-1}\left(\frac{\omega_n L - \frac{1}{\omega_n C}}{R}\right). \qquad (2.116)$$

Since every term in Eq. (2.111) has the assumed form, our formal solution for a drive of the form $f(t)$ is then

$$I = \sum_{n=-\infty}^{\infty} \hat{I}_n. \qquad (2.117)$$

Fourier's theorem means that we can think of any periodic function as being composed of oscillating terms at different frequencies, and that these frequencies

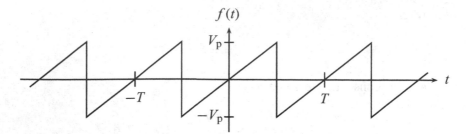

Figure 2.28 A symmetric sawtooth.

are harmonically related (i.e., the higher frequencies are integer multiples of the first frequency ω_1, as noted in Eq. (2.112)). This is useful because, as we have seen with the various filters, circuit response is often a function of frequency. Thus, for example, a periodic function $f(t)$ applied to a low-pass filter may have its lower frequency components unaffected while its higher frequencies are attenuated. The output would be roughly the sum of the lower frequency portions of the original signal.

The only difficult part of using Fourier's theorem is computing the constants $\hat{c}_n$ using Eq. (2.113) and simplifying the resulting series. This is not usually required in electronics since the typical periodic functions used have their Fourier series already tabulated, but for the curious we do one example. Suppose our periodic function is the sawtooth wave shown in Fig. 2.28.

Since the integral in Eq. (2.113) can be evaluated for any beginning time t', we choose $t' = -\frac{T}{2}$ for convenience. Then over the limits of the integral ($-\frac{T}{2}$ to $\frac{T}{2}$), the function can be expressed as $f(t) = (2V_p/T)t$. Using this in Eq. (2.113) gives

$$\hat{c}_n = \frac{1}{T} \int_{-\frac{T}{2}}^{\frac{T}{2}} dt \frac{2V_p}{T} t e^{-j\omega_n t}$$

$$= \frac{2V_p}{T^2} \left[\frac{t e^{-j\omega_n t}}{-j\omega_n} \Bigg|_{-\frac{T}{2}}^{\frac{T}{2}} - \frac{1}{-j\omega_n} \int_{-\frac{T}{2}}^{\frac{T}{2}} dt e^{-j\omega_n t} \right] \quad \text{(for } n \neq 0\text{)} \quad (2.118)$$

$$= 0 \quad \text{(for } n = 0\text{)}$$

where for the $n \neq 0$ case we have integrated by parts. The first term in Eq. (2.118) simplifies:

$$t e^{-j\omega_n t} \Big|_{-\frac{T}{2}}^{\frac{T}{2}} = \frac{T}{2} e^{-jn\pi} - \left(-\frac{T}{2} \right) e^{jn\pi} = T \cos n\pi = T(-1)^n. \quad (2.119)$$

So does the second term:

$$\int_{-\frac{T}{2}}^{\frac{T}{2}} dt\, e^{-j\omega_n t} = \frac{1}{-j\omega_n} e^{-j\omega_n t}\Big|_{-\frac{T}{2}}^{\frac{T}{2}} = -\frac{1}{j\omega_n}(e^{-jn\pi} - e^{jn\pi}) = \frac{2}{\omega_n}\sin n\pi = 0.$$

$$(2.120)$$

We are thus left with

$$\hat{c}_n = -\frac{2V_p}{j\omega_n T}(-1)^n \tag{2.121}$$

for all n except zero. Our sawtooth function can then be written as

$$f(t) = \sum_{n=-\infty}^{\infty} -\frac{2V_p}{j\omega_n T}(-1)^n e^{j\omega_n t}$$

$$= \sum_{n=1}^{\infty} \frac{2V_p}{T}(-1)^n \frac{j}{\omega_n}\left(e^{j\omega_n t} - e^{-j\omega_n t}\right)$$

$$= \sum_{n=1}^{\infty} \frac{2V_p}{\pi}(-1)^{n+1}\frac{1}{n}\sin\omega_n t. \tag{2.122}$$

Writing out the series and noting $\omega_n = n\omega_1$ yields

$$f(t) = \frac{2V_p}{\pi}\left[\sin\omega_1 t - \frac{1}{2}\sin 2\omega_1 t + \frac{1}{3}\sin 3\omega_1 t - \frac{1}{4}\sin 4\omega_1 t + \cdots\right]. \tag{2.123}$$

Our Fourier analysis has thus shown us that the sawtooth wave can be viewed as consisting of a series of sine waves with frequencies that are integer multiples of $\omega_1 = \frac{2\pi}{T}$. Similar series can be written for other common waveforms. For a triangle wave with peak amplitude V_p and period T we obtain

$$f(t) = \frac{8V_p}{\pi^2}\left[\sin\omega_1 t + \frac{1}{9}\sin 3\omega_1 t + \frac{1}{25}\sin 5\omega_1 t + \frac{1}{49}\sin 7\omega_1 t + \cdots\right] \tag{2.124}$$

while for a square wave with the same specifications we have

$$f(t) = \frac{4V_p}{\pi}\left[\sin\omega_1 t + \frac{1}{3}\sin 3\omega_1 t + \frac{1}{5}\sin 5\omega_1 t + \frac{1}{7}\sin 7\omega_1 t + \cdots\right]. \tag{2.125}$$

2.9 Transformers

A transformer is, as the name implies, a device that *transforms* an AC voltage of one amplitude into an AC voltage of another amplitude. This ability is routinely

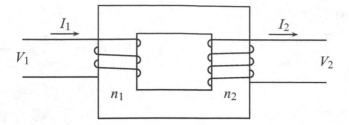

Figure 2.29 Schematic of a transformer.

used throughout the electrical systems of modern society. At power generating plants, huge transformers step up the voltage so that a large amount of power can be transmitted without necessitating very large transmission wires. Since the high voltages are dangerous, at the consumer end of the power grid transformers step down the voltage to a safe level. The required voltage supplies inside most consumer electronics are even lower, so another step-down transformer is found inside these devices.

A typical transformer is shown schematically in Fig. 2.29. A length of wire is wrapped around one portion of the transformer core n_1 times. These windings are called the *primary* windings. Another wire is wrapped around the core n_2 times to form the *secondary* windings. An AC voltage V_1 is applied to the primary coil and an AC current I_1 flows. This produces a time-varying magnetic field. The core is made of a ferrous material so that the magnetic field tends to stay inside the material and is guided around to where the secondary coil is positioned. Faraday's Law tells us that a time-varying magnetic field inside a coil of wire induces a voltage in that coil, and we will call that secondary voltage V_2. The relationship between V_1 and V_2 is expressed by

$$V_2 = \left(\frac{n_2}{n_1}\right) V_1. \tag{2.126}$$

The output voltage of the transformer V_2 thus depends only on the input voltage V_1 and the ratio of the number of secondary to primary turns, n_2/n_1.

If a resistor or other load is attached to the secondary coil, a current I_2 can flow. Since energy must be conserved (we assume no losses due to flux leakage or other non-ideal behaviors), $I_1 V_1 = I_2 V_2$. Using Eq. (2.129) for V_2, we obtain

$$I_2 = \left(\frac{n_1}{n_2}\right) I_1. \tag{2.127}$$

The output current of the transformer I_2 depends only on the input current I_1 and the so-called *turns ratio*, n_1/n_2. Note that if we produce a transformer with a turns

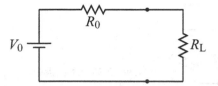

Figure 2.30 The power transfer problem.

ratio that will increase the voltage, the output current will necessarily be less than the input current, and vice versa. There is no free lunch in physics.

Equations (2.126) and (2.127) are the fundamental laws of ideal transformer operation, and these are the basis of the step-up and step-down transformers mentioned earlier. Another use for the transformer is in *impedance matching*. To establish the motivation for this usage, let's consider the following problem. Suppose we have a voltage source with a voltage V_0 and an internal resistance R_0. This is connected to a load resistor R_L as shown in Fig. 2.30. The question is, how should R_0 and R_L be related if we want the maximum power transferred to the load resistor?

The power to the load is

$$P_L = I^2 R_L = \left(\frac{V_0}{R_0 + R_L} \right)^2 R_L. \tag{2.128}$$

Note that the power to the load goes to zero when $R_L \to 0$ or $R_L \to \infty$. Somewhere between these limits we will get a maximum in the power. To find this case we set the derivative of P_L with respect to R_L to zero:

$$\frac{dP_L}{dR_L} = V_0^2 \left[\frac{1}{(R_0 + R_L)^2} - \frac{2R_L}{(R_0 + R_L)^3} \right] = 0. \tag{2.129}$$

Solving this for R_L gives $R_L = R_0$ for maximum power transfer.

This general result shows that the load resistance (or, more generally, the load impedance) must match the internal resistance of the source if we want to transfer the most power to the load. While this is desirable, we are sometimes faced with situations where this is not the case. For example, most stereo speakers have an impedance of 8 Ω and this load impedance may not match the output impedance of the stereo amplifier. To achieve the desired impedance matching, we can employ a transformer. Consider the circuit shown in Fig. 2.31.

The output current and voltage are given by $V_2 = I_2 R_L$. Using Eqs. (2.126) and (2.127) for V_2 and I_2 we get

$$V_1 \left(\frac{n_2}{n_1} \right) = I_1 \left(\frac{n_1}{n_2} \right) R_L \tag{2.130}$$

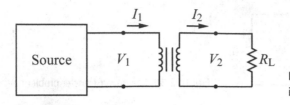

Figure 2.31 A transformer used to match impedance.

or

$$V_1 = I_1 \left[\left(\frac{n_1}{n_2} \right)^2 R_L \right]. \tag{2.131}$$

Thus the input voltage and current are now related by an effective resistance

$$R_{\text{eff}} = \left(\frac{n_1}{n_2} \right)^2 R_L. \tag{2.132}$$

The point here is that the transformer has changed the resistance the source "sees" from R_L to R_{eff}. By choosing a transformer with appropriate turns ratio, we can match R_{eff} to whatever source resistance we have, thereby insuring maximum power transfer. In effect, the transformer has matched the source impedance to the load impedance.

Further details on analyzing circuits involving transformers are given in Appendix C.

EXERCISES

1. Find the equivalent capacitance across the terminals of the circuit in Fig. 2.32.

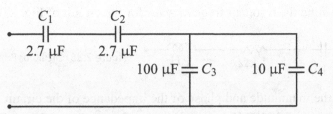

Figure 2.32 Circuit for Problem 1.

2. Sketch V_{out} versus time for the circuit shown in Fig. 2.33 after the switch is closed. Assume the capacitor is initially uncharged and that $V_0 = 12\,\text{V}$, $R = 100\,\text{k}\Omega$, and $C = 10\,\mu\text{F}$. Include appropriate numeric scales on the axes.

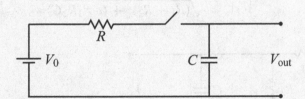

Figure 2.33 Circuit for Problems 2, 3, and 4.

3. Referring to Fig. 2.33 and taking $V_0 = 100\,\text{V}$, $R = 1\,\text{M}\Omega$, what value of C is needed so that $V_{out} = 70\,\text{V}$ at 10.0 s after the switch closes?

4. Suppose we use the circuit of the previous problem and wait a long time after the switch is closed so that the capacitor is fully charged. Now we open the switch and attach a 10 kΩ resistor across the output terminals. How long will it take for the voltage across the capacitor to drop to 1.0 V?

5. A sine wave with amplitude 20 V_{pp} is connected to a 10 kΩ resistor. Calculate the peak, the rms, and the average currents though the resistor. What power rating should the resistor have?

6. Calculate the magnitude and the phase of the total impedance for the circuit shown in Fig. 2.34.

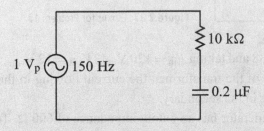

Figure 2.34 Circuit for Problems 6 and 7.

7. Suppose we change the frequency of the signal generator in Fig. 2.34. If the angular frequency is set to 10^3 rad/s, what is the peak amplitude of the voltage across the capacitor? Of the voltage across the resistor?

8. Determine the resonant frequency ω_0 for the circuit of Fig. 2.35.

2 µF   1 kΩ   1 H   **Figure 2.35** Circuit for Problems 8 and 9.

9. Find the magnitude and phase of the impedance of the circuit of Fig. 2.35 for a frequency of 2500 Hz.

10. Design an RC low-pass filter that has $|V_{out}|/|V_{in}| = 0.5$ at 5 kHz.

11. Derive the following expression for the circuit of Fig. 2.36:

$$\frac{|V_{out}|}{|V_s|} = \frac{R_2}{\sqrt{(R_1 + R_2)^2 + (\omega R_1 R_2 C)^2}}. \tag{2.133}$$

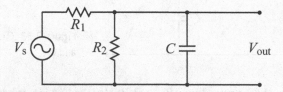

Figure 2.36 Circuit for Problem 11.

12. Consider the circuit of Fig. 2.37, where $R_1 = 20\,\Omega$, $\chi_1 = 37.7\,\Omega$, $R_2 = 10\,\Omega$, and $\chi_2 = -53.1\,\Omega$. Compute the magnitude of the current flowing out of the signal generator and the phase angle between that current and the signal generator voltage. Assume the signal generator outputs a 230 V_{rms} sine wave with a frequency of 60 Hz.

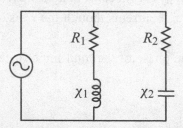

Figure 2.37 Circuit for Problem 12.

13. Refering to Fig. 2.38 and taking $V_{in} = 120\,V_{rms}$, $V_{out} = 12\,V_{rms}$, and $R = 20\,\Omega$, find the turns ratio of the transformer, the current flowing in the primary, and the current flowing in the secondary.

14. An audio signal generator has an output impedance of 600 Ω. To drive an 8 Ω speaker with maximum power transfer, an impedance-matching transformer is

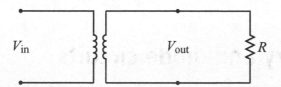

Figure 2.38 Circuit for Problem 13.

used between the generator and the speaker. What is the necessary turns ratio for such a transformer?

15. A step-down transformer with no markings is measured to have a $1.6\,\Omega$ impedance on one side and a $40\,\Omega$ impedance on the other. Which side is the primary and which side is the secondary? If the input voltage is $120\,V_{rms}$, what will the output voltage be?

16. A low-pass filter with a break-point frequency of 100 Hz is used to filter a 90 Hz square wave signal with amplitude V_0. Describe the output of the filter including its rough shape, its frequency, and its amplitude.

17. Write down the Fourier series for the signal shown in Fig. 2.39. Hint: it is not necessary to do any calculation. Think about how this signal is related to one for which we already know the Fourier series.

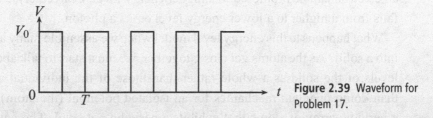

Figure 2.39 Waveform for Problem 17.

18. Given a sawtooth signal with a period of 1 ms, design a circuit (including all component values) that will take this sawtooth as input and give a 3 kHz sine wave as output. If the peak voltage of the sawtooth is 1 V, what will the amplitude of the resulting sine wave be?

FURTHER READING

Charles K. Alexander and Matthew N. O. Sadiku, *Fundamentals of Electric Circuits*, 2nd edition (New York: McGraw-Hill, 2004).

L. W. Anderson and W. W. Beeman, *Electric Circuits and Modern Electronics* (New York: Holt, Rinehart, and Winston, 1973).

A. James Diefenderfer and Brian E. Holton, *Principles of Electronic Instrumentation*, 3rd edition (Philadelphia, PA: Saunders, 1994).

Robert E. Simpson, *Introductory Electronics for Scientists and Engineers*, 2nd edition (Boston, MA: Allyn and Bacon, 1987).

3 Band theory and diode circuits

3.1 The band theory of solids

3.1.1 Introduction

A fundamental result from basic modern physics is that atoms are characterized by discrete energy levels. Each of these energy levels can accept up to two electrons. When "building" an atom, we start from the lowest level, fill in two electrons, and then move up to the next energy level and fill it with electrons. This continues until we have placed all the atom's electrons in energy levels. We also know that if an atom absorbs energy from the outside (for example, by absorbing a photon), an electron can be promoted to a higher energy level. Conversely, an electron that falls from a higher to a lower energy level emits a photon.

What happens to this energy level model when we assemble many atoms together into a solid? As the atoms get closer together, we must start to talk about the energy levels of the solid as a whole rather than those of the individual atoms. Rather than doing quantum mechanics for an isolated potential (the atom), we do it for a periodic array of atoms that exhibits a periodic potential. The net result of this is that, during the assembly of N atoms, the individual atomic levels split into N levels. This is shown schematically in Fig. 3.1. Thus when the solid is assembled and the atoms are at their final equilibrium spacing, the solid is characterized by a series of *energy bands* consisting of a large number of closely spaced allowed energy levels. Just as electrons in individual atoms cannot have energies between the atomic energy levels, so electrons in a solid are forbidden to have energies between the allowed bands.

3.1.2 Band theory

Imagine we are constructing a material consisting of N atoms as described above, but we have not added any electrons yet. We now start to add electrons to the solid, starting with the lowest energy level. As with atomic energy levels, only certain

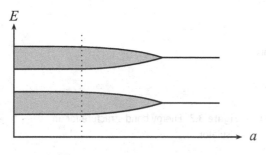

Figure 3.1 Schematic representation of energy level splitting as a solid is assembled. Here *a* is the distance between atoms. The light gray shading represents the large number of closely spaced energy levels produced by the splitting of the original atomic energy level. The dotted line represents the equilibrium spacing. At this spacing the solid is characterized by energy bands.

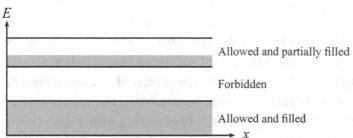

Allowed and partially filled

Forbidden

Allowed and filled

Figure 3.2 Energy band structure for a conductor.

energies are allowed and we can only place two electrons in each level due to the Pauli Exclusion Principle. We continue to fill energy levels from the bottom up until we have used all the electrons normally associated with our N atoms. The character of the resulting band structure determines whether a material is a conductor, an insulator, or a semiconductor.

Consider, for example, the case shown in Fig. 3.2, where we plot energy level versus position x within the material. We show only the upper energy levels since these are the ones of interest; the lower energy levels are all filled with electrons and do not change under normal material operations. Moving from the bottom of the figure up, we show a band of allowed energy levels that are filled with electrons, followed by a band of non-allowed (or forbidden) energies, followed by an allowed band that is only partially filled with electrons.

It is the fact that this last band is only partially filled that makes this material a *conductor*. In order to produce a current, electrons in a material must move and thus must increase their energy slightly. They must, therefore, be able to move to a slightly higher energy level. This is possible for this material because there are lots of empty energy levels in the top-most band. This material will thus be a good conductor.

This situation should be contrasted with the band structure shown in Fig. 3.3. Here we have a full, allowed band, followed by a wide forbidden band, followed by an empty allowed band. The electrons in this material cannot flow to produce a current because there are no nearby unfilled energy levels for them to move to.

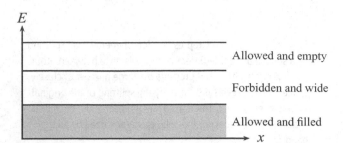

Figure 3.3 Energy band structure for an insulator.

There are unfilled levels further up, but the forbidden band is too wide for the electrons to cross. This is the characteristic band structure of an *insulator*.

At this point we should quantify some of these statements. The energy gained by an electron contributing to a current in a material is quite small. The voltage applied to produce the current accelerates the electrons, but before they gain much velocity they undergo a collision that changes their direction and slows them down. As a result, the average drift energy of a conduction electron is on the order of 10^{-21} eV. This is comparable to the spacing between energy levels within a band, but is much smaller than the energy separation between bands (~ 0.1 eV). Thus there is no way the electrons in an insulator can gain enough energy from conduction to cross the forbidden band.

There is, however, another source of energy that plays an important role in semiconductor physics: thermal energy. A basic insight from thermal physics tells us that temperature is a macroscopic measure of the microscopic motion of particles. Thus particles in our material can have energies beyond those set by their position on the band structure. The additional energy from thermal motion has a distribution of values (more on this later), but is of order kT, where k is Boltzmann's constant. At room temperature, $kT \approx 0.025$ eV, which is comparable to the energy gap between bands. It is thus possible (if the temperature is high enough and the band gap is small enough) for an electron to jump to the next allowed band. Once it is there, it can produce a current because now there are lots of the open energy levels required for conduction. If our material is indeed an insulator, the forbidden band must be wide enough to prevent this from happening, i.e., it must be too wide for so-called thermal transitions.

Our third type of material, the *semiconductor*, is characterized by the opposite situation: the forbidden band is narrow enough to allow thermal transitions to the next allowed band. This is shown in Fig. 3.4. As with the insulator, we have a filled allowed band, followed by a forbidden band and an empty allowed band. If the temperature of the material is non-zero, the electrons can move into the

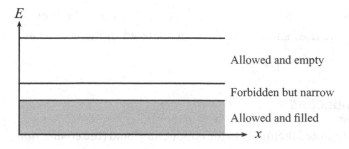

Figure 3.4 Energy band structure for a semiconductor.

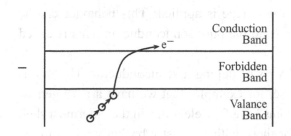

Figure 3.5 Schematic depiction of hole formation and transport when an electron is promoted.

empty allowed band using thermal energy. We thus see on a basic level why semiconductor electronics is sensitive to temperature: the temperature determines how much thermal energy is available to promote electrons from the last filled band (called the *valance band*) up to the next allowed band (called the *conduction band*).

When an electron in a semiconductor moves from the valance band to the conduction band, it leaves behind a vacancy in the valance band. This vacancy is called a *hole* (see Fig. 3.5). The hole behaves as if it were a positive particle. To see this, imagine that a voltage is applied across our material, with the negative terminal of the voltage on the left and the positive terminal of the voltage on the right. If an electron is promoted to the conduction band, it will also move toward the right, toward the positive voltage. The hole it left behind in the valance band will soon be filled with an electron in a lower energy level. While moving up in energy, this electron will also move to the right, trying to get to the positive terminal. Thus the vacancy it leaves will move to the left. As this process of filling the vacancy continues, the hole will move down and to the left, toward the negative terminal. It is thus possible to view the process in terms of the motion of the hole rather than the motion of the electrons. In this view, the hole (a positive conduction particle) moves toward the negative terminal as it increases in energy (here we must switch signs on our energy scale and let the hole energy increase downwards). If all this seems confusing, don't worry: the hole picture is not required. We can still get the physics right by just following the motion of the electrons, and this is the approach we will take in this text. We include the hole picture for completeness, since it is often encountered in the electronics literature.

3.1.3 Doping semiconductors

The number of electrons promoted thermally to the conduction band (the *conduction electrons*) is small for a pure semiconductor, so a pure semiconductor will only allow a small current to flow when a voltage is applied. This behavior can be altered, however, by adding impurities to the pure semiconductor. This is called *doping* the semiconductor.

For our purposes, there are two ways of doping a semiconductor. The first is by adding a *donor impurity*. Suppose, for example, that we have a pure germanium semiconductor. Germanium atoms have four electrons in the outermost shell of the atom (the outermost shell, for those with chemistry background, is in the $4s^2 4p^2$ configuration). If this is doped with antimony (which has five electrons in the outermost shell $5s^2 5p^3$), the semiconductor then has a number of extra loosely bound electrons. The effect of this is to add filled, localized energy levels to the band structure (*filled* with the extra electrons and *local* because the impurities are in particular positions within the semiconductor). The location of these additional levels depends on the type of material, type of impurity, and other factors. A desirable location is shown in Fig. 3.6. The new levels are near the bottom of the conduction band. It is thus relatively easy for these electrons to be promoted into the conduction band compared to those in the valance band of the pure semiconductor. Thus a donor impurity makes it easier for the semiconductor to conduct. Because the charges that produce this current (the so-called charge carriers) are electrons, this type of doped semiconductor is called an n-type semiconductor (n for *negative*).

Conversely, suppose we dope our pure germanium semiconductor with gallium (outermost shell $4s^2 4p^1$). Gallium has one less electron than germanium and thus

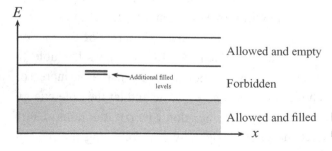

Figure 3.6 Schematic representation of an n-type semiconductor.

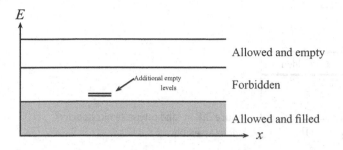

Figure 3.7 Schematic representation of a p-type semiconductor.

has a vacancy in its highest energy level. Such an impurity is called an *acceptor*. The effect of an acceptor impurity is to add empty, localized energy levels to the band structure. In this case, it is desirable to have these additional energy levels located just above the valance band, as shown in Fig. 3.7. It is then easy for electrons from the valance band to move up into these empty levels. But since these levels are localized, the promoted electrons cannot move through the material as they must in order to contribute to the current. The holes left behind in the valence band, on the other hand, can move throughout the material since the valence band energy levels are not localized. We thus can again have an enhanced level of current due to the impurity, but now the current is produced by the *positive* hole motion. This type of doped semiconductor is thus called p-type.

Note that both n-type and p-type semiconductors are electrically neutral just as the atoms that make up the material are electrically neutral; the "n" does not stand for negatively charged, but for negative charge carriers, and similarly for the "p." In addition, although the current is carried predominately by electrons in the n-type semiconductor, there are still a few holes produced by electrons that are promoted from the valence band to the conduction band, so a small portion of the current is carried by holes, even in the n-type materials. The electrons in the n-type material are called *majority charge carriers*, while the holes are called *minority charge carriers*. Conversely, in p-type material, the current is predominately carried by the hole motion, but a few electrons promoted from the valence band to the conduction band also contribute, so the holes in a p-type material are the majority charge carriers while the electrons are the minority charge carriers.

3.1.4 The p-n junction

Now we consider what happens when we bring together a piece of n-type semi-conductor and a piece of p-type semiconductor. We represent our two pieces as shown in Fig. 3.8. The n-type material has more electrons in its conduction band

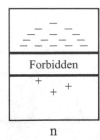
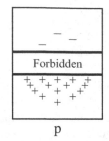

Figure 3.8 n- and p-type semiconductors before assembly.

and the p-type material has more holes in the valance band. For the n-type material, the number of electrons in the conduction band decreases with energy because it is harder for the electrons to be promoted to these higher levels (we will quantify this in a moment). Similarly, there are fewer holes at the higher hole energy levels (remember, hole energy increases *downward* in this diagram).

When the two materials are brought together, the higher density electrons in the n-material will diffuse into the p-material where the density of electrons in the conduction band is lower. The opposite happens for the high density holes in the p-material; they diffuse into the n-material. (Because the behavior of electrons and holes is analogous, from now on we will focus our attention on the electrons alone.) This diffusion of charges leads to a charge imbalance, with excess electrons accumulating in the p-material. As this continues, an electric field builds up at the junction pointing from the n- to the p-material, and this field opposes further diffusion of electrons. In the end, the field adjusts itself so that an equilibrium is established and the flow of electrons from n- to p-material is the same as that from p- to n-material. The region near the p-n junction where this diffusion takes place is called the *depletion region* because the density of the charge carriers (electrons in the n-material and holes in the p-material) is markedly reduced.

What does all this do to the band structure? The electric field E at the junction requires a step in the electrostatic potential V (recall $E = -dV/dx$) and this produces an opposite step in the electron energy (since energy $E = qV$, and q for an electron is negative). Thus, in equilibrium, the band structure of the p-n junction will appear as shown in Fig. 3.9. The energy levels shift up by an amount ΔE going from the n-material to the p-material. The magnitude of ΔE will be just enough to insure that the flow of electrons in each direction (f_1 and f_2 in the figure) is balanced. Electrons in the p-material that make their way to the junction have no trouble moving into the n-material and produce an electron flow f_2. Similarly, electrons in the n-material that make their way to the junction produce an electron flow f_1 into the p-material. Originally, f_1 is greater than f_2 because there are more electrons on the n-side than on the p-side. The energy band shift, however, prevents some of

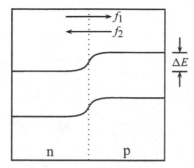

Figure 3.9 An unbiased p-n junction.

the n-material electrons from moving into the p-material because this would place them in the forbidden band of the p-material (note that the electrons must move straight across the diagram since their energy stays constant). Only those n-side electrons with energy higher than the top of the p-material forbidden band can move into the p-material, and by adjusting ΔE this flow can be made to match f_2.

To quantify this, we note that the electrons in the conduction band have been promoted to that level by thermal energy. A basic result from thermal physics says that the density of particles F with energy E in a system at temperature T is given by

$$F = Ae^{-E/kT} \tag{3.1}$$

where A is a normalization constant and k is Boltzmann's constant.[1] Then the number of particles N (here, electrons) with energy above a level $E_0 + \Delta E$ is then

$$N = A \int_{E_0+\Delta E}^{\infty} e^{-E/kT}\, dE = -AkT \left(e^{-\infty} - e^{-(E_0+\Delta E)/kT} \right) = AkT e^{-E_0} e^{-\Delta E/kT}. \tag{3.2}$$

Here E_0 would correspond to the energy at the bottom of the conduction band. Since the flow of electrons f_1 will be proportional to this number, we get

$$f_1 = Ce^{-\Delta E/kT} \tag{3.3}$$

where we have grouped all the constants into C. Since, in equilibrium, there is no net flow of electrons, we must also have

$$f_2 = Ce^{-\Delta E/kT}. \tag{3.4}$$

---

[1] This is only approximately correct. A more rigorous approach would use the Fermi–Dirac distribution for F.

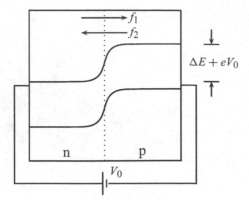

Figure 3.10 Reverse bias of the p-n junction.

Now let's examine what happens when we apply an external voltage V_0 to our p-n junction. If the negative side of our voltage is applied to the p-material and the positive side to the n-material, we get the situation shown in Fig. 3.10. The energy levels on the p-side have been raised by an amount eV_0, where e is the charge of an electron. This will make it harder for electrons to flow from the n- to the p-material, and this is consistent with the notion that electrons will tend to be repelled from the negative bias on the p-side. Quantifying this effect, we see that the flow of particles from the n-side to the p-side will now be

$$f_1 = Ce^{-(\Delta E + eV_0)/kT}. \tag{3.5}$$

The shift in energy levels will have no effect, however, on the flow f_2 of electrons from the p- to the n-side, so this remains the same as given in Eq. (3.4). There is thus now a net flow of electrons of

$$f_{net} = f_1 - f_2 = Ce^{-\Delta E/kT}\left(e^{-eV_0/kT} - 1\right). \tag{3.6}$$

For room temperature, $kT \approx 0.025$ eV, so for any appreciable voltage V_0, the term $\exp\left(-eV_0/kT\right) \rightarrow 0$ and we have

$$f_{net} \approx -Ce^{-\Delta E/kT}. \tag{3.7}$$

Note that for this case (called the *reverse biased* case), the net flow is independent of applied voltage V_0 but strongly dependent on temperature (cf. Eq. (3.7)). Also, since the energy step in the band structure is larger than in the equilibrium case, more charge carriers are required to support it and the depletion layer is widened.

Now we apply an external voltage V_0 to our p-n junction such that the negative side of our voltage is applied to the n-material and the positive side to the p-material. This is called the *forward biased* case. For forward bias, we get the situation shown

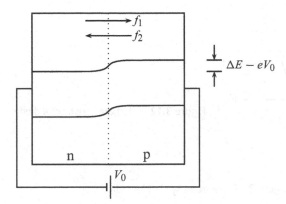

Figure 3.11 Forward bias of the p-n junction.

in Fig. 3.11. The step in the energy levels is now reduced to $\Delta E - eV_0$, and this will make it easier for electrons to flow from the n- to the p-material. This, again, is consistent with the notion that electrons will tend to be attracted to the positive bias on the p-side. As before, f_2 is unaffected by the bias, but f_1 changes to

$$f_1 = Ce^{-(\Delta E - eV_0)/kT} \tag{3.8}$$

and the net flow becomes

$$f_{\text{net}} = f_1 - f_2 = Ce^{-\Delta E/kT}\left(e^{eV_0/kT} - 1\right). \tag{3.9}$$

Thus, for any appreciable forward bias V_0, the first term in parentheses of Eq. (3.9) dominates and we have an exponential increase in the flow of electrons from the n-side to the p-side. Since the energy step in the band structure is smaller than in the equilibrium case, the depletion layer is narrowed.

We can now combine these two cases by defining the forward bias as a positive applied voltage V_d. We also return to using current to describe the flow of charge, noting that the current will be in the opposite direction of the net electron flow. The behavior of our p-n junction (called a *diode*) can then be summarized as

$$I = I_0\left(e^{eV_d/kT} - 1\right) \tag{3.10}$$

where $I_0 \equiv C\exp\left(-\Delta E/kT\right)$. A graph of this result is shown in Fig. 3.12. I_0 is usually very small compared to a typical forward bias current and is thus often approximated as zero.

The electronic symbol for this device is shown in Fig. 3.13 along with the bias polarity and current direction for forward biased operation. We have also indicated which end of the diode corresponds to the n- and p-material. Note that the filled triangle of the symbol points in the direction of the forward biased current. The

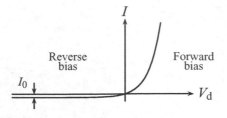

Figure 3.12 *I–V* characteristic of a diode.

Figure 3.13 Electronic symbol for a diode.

n-side of the diode is sometimes referred to as the *cathode* while the p-side is called the *anode*.

3.1.5 Breakdown

If the p-n junction is strongly reverse biased, $f_1 \approx 0$ and f_2 is limited by the presence of electrons in the p-material conduction band due to thermal excitation. At some point, however, the increasing electric field in the depletion layer causes two types of breakdown phenomena which strongly increase the reverse current.

1. Avalanche breakdown. In this type of breakdown, electrons from the p-side are accelerated to high enough kinetic energy to ionize other atoms in the depletion layer, thus producing a new electron-hole pair. The new electron is also accelerated and can produce more pairs, and so on. The resulting chain reaction adds many electrons to the conduction band and thus rapidly increases the current.
2. Zener breakdown. In this case, the electric field in the depletion layer becomes large enough to produce ionization directly, essentially tearing the atoms apart. This process again produces copious electron-hole pairs and rapidly increases the current.

   This effect modifies the diode *I–V* characteristic as shown in Fig. 3.14. Note that while the breakdown current increases rapidly, the voltage stays fairly constant. The magnitude of this voltage is called the *breakdown voltage*. Despite the name, both types of breakdown are non-destructive. That is, the diode can be operated in breakdown mode without destroying it, and, as we shall see later, certain circuits deliberately use the steep rise of the reverse current to achieve useful results.

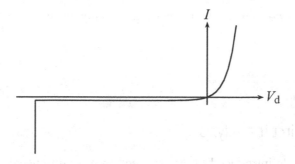

Figure 3.14 Diode I–V characteristic showing breakdown at large reverse bias.

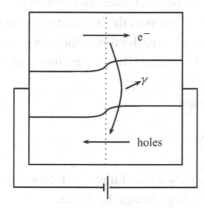

Figure 3.15 Photon emission due to electron-hole recombination.

3.1.6 Photon emission and absorption

Another interesting consequence of our band theory model is the possibility of producing light from a p-n junction or changing the electrical properties of the junction through photon absorption (Fig. 3.15). Recall that, for an atom, a photon is emitted when an electron moves from a higher energy state to a lower one. A similar phenomenon can occur in our p-n junction. When the diode is forward biased, lots of electrons and holes flow in opposite directions through the depletion region. Since the holes represent vacancies in a lower energy level, it is possible for the electron to jump to one of these lower energy levels while a photon is emitted to conserve energy. If the diode is constructed so that these photons can exit the material, we have a light source. This is the basis of the ubiquitous *light emitting diode* or *LED*, which is used as an indicator light on many modern electronic devices.

Photons can also be *absorbed* by our p-n junction. In this case, an incoming photon promotes an electron from the valance band to the conduction band, thus producing a new electron-hole pair. If these new pairs are produced in significant numbers, they can significantly alter the ability of the junction to conduct electricity.

This is the basis of the *photo-diode*, which is used in many applications as a light detector.

3.2 Diode circuits

3.2.1 Basic diode circuit analysis

The semiconductor diode introduced in the last section has an unusual I–V characteristic: it conducts readily for forward bias but does not conduct very much for reverse bias (assuming the reverse bias is less than the breakdown voltage). As we will see, this unusual behavior allows us to use the diode for many purposes, but it also complicates the analysis of diode circuits. To see why this is true, consider the simple diode circuit shown in Fig. 3.16.

Applying KVL to this circuit we obtain

$$V_0 - IR_{\mathrm{L}} - V_{\mathrm{d}} = 0 \tag{3.11}$$

where V_{d} is the voltage across the diode. We also know the relationship between the current through the diode and the voltage across the diode:

$$I = I_0 \left(e^{eV_{\mathrm{d}}/kT} - 1 \right). \tag{3.12}$$

Solving Eq. (3.11) for I and combining with Eq. (3.12) yields

$$I = \frac{V_0 - V_{\mathrm{d}}}{R_{\mathrm{L}}} = I_0 \left(e^{eV_{\mathrm{d}}/kT} - 1 \right). \tag{3.13}$$

In this last equation, the only unknown is V_{d}. If we knew this, we could plug into any of the other equations and obtain I and we would be done. But Eq. (3.13) is a transcendental equation and cannot be solved analytically for V_{d}, and herein is the complicating factor in the analysis of diode circuits.

There are two standard ways of dealing with transcendental equations. One is to solve the equation numerically using a technique such as Newton's Method.

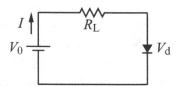

Figure 3.16 Simple diode circuit.

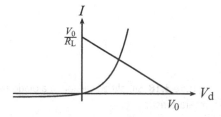

Figure 3.17 Diode I–V characteristic and load line for the circuit of Fig. 3.16.

A second way is to solve the equation graphically, and this will help us to see some of the key features of the solution. To proceed, we note that Eq. (3.13) gives two equations for I as a function of V_d, so we plot both of these on the same graph. The intersection of the two curves tells us the value of V_d (and thus I) that satisfies both equations. In electronics, this solution is called the *operating point*.

This procedure is shown in Fig. 3.17. The diode I–V characteristic (Eq. (3.12)) is plotted along with the linear equation $I = (V_0 - V_d)/R_L$. This latter equation is referred to as the *load line*, and this graphical solution is often called the *load line method* in electronics. Note that the x- and y-intercepts for the load line are V_0 and V_0/R_L, respectively, and the slope of the line is $-R_L^{-1}$. We can thus imagine what would happen to the solution if V_0 or R_L was varied. For example, if V_0 is varied, both intercept points will move along their respective axes while the slope of the line stays fixed. If V_0 increases, the operating point will move up the diode characteristic and the current will rapidly increase. Also note that the analysis is not restricted to positive V_0. If V_0 is negative, the x- and y-intercepts will be on the negative portion of the x- and y-axes and the operating point will be on the relatively constant, reverse biased portion of the diode characteristic, thus showing that little current flows in the circuit for this case.

While the load line method offers some insights into the detailed behavior of the circuit, it is cumbersome to use for routine circuit analysis. It is thus common practice to employ a simplified model of the diode I–V characteristic that allows for analytical solutions. In this text, we will use the simplified characteristic shown in Fig. 3.18. For $V_d < 0.6$ V, the diode current is zero. At $V_d = 0.6$ V, the characteristic becomes a vertical line. This simplified model keeps two of the important features of the real characteristic curve: the current rises rapidly for $V_d \approx 0.6$ V and is small otherwise. Our simplified model may also be expressed in words: the diode will not allow current flow unless it is forward biased; when it is forward biased, the voltage drop across the diode is 0.6 V.[2]

---

[2] The value 0.6 V is appropriate for diodes made from silicon, the most common material.

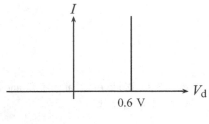

Figure 3.18 Simplified version of a diode I–V characteristic.

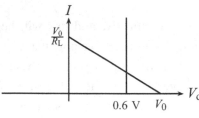

Figure 3.19 Load line plotted on a simplified diode I–V characteristic.

To see how this simplifies our circuit analysis, let's return to the circuit of Fig. 3.16. Assuming the diode is forward biased, Eq. (3.11) now yields

$$I = \frac{V_0 - 0.6}{R_L} \qquad (3.14)$$

where we have used our approximation $V_d = 0.6$ V. Since we know V_0 and R_L, we can obtain the circuit current I directly without the use of a graphical or numerical method.

Several caveats are now in order. We have used the approximation $V_d = 0.6$ V for the forward biased diode. Two other common approximations are often seen in textbooks. The first simply uses $V_d = 0.7$ V for the diode voltage. Since the actual diode voltage depends on the current, it is hard to argue persuasively for either value, so we simply note the difference in convention. On the other hand, some textbooks use $V_d = 0$. While this makes analysis even easier, in this case the approximation loses important information: a forward biased diode has a non-zero voltage drop. Without this understanding, some laboratory observations will be puzzling and some electronic circuits will seem without merit (we shall see some examples later).

Another problem lies in our assumption of the diode being forward biased. When we are doing circuit analysis, how do we know at the start if the diode will be forward biased? What do we do if it is not? To answer these questions, we return to the load line analysis. Figure 3.19 shows our simplified diode characteristic plotted along with the load line. From this we see that in order to be forward biased (and thus have an operating point on the vertical part of the diode characteristic), we must have $V_0 > 0.6$ V. If this is not true, the load line will cross the diode characteristic where $I = 0$ and give an operating point with $V_d = V_0$. This is consistent with our general voltage loop law notions: if the current in the circuit is zero, there is no voltage drop across the resistor so the voltage across the diode must be equal to the battery voltage.

An alternative tactic to use to avoid mistakes is to check your answer for consistency. Suppose our circuit specified $V_0 = 0.4$ V. We know from the previous paragraph that this is not enough to forward bias the diode, but suppose we did not know this and made the approximation $V_d = 0.6$ V. We proceed as before and get Eq. (3.14), but plugging in V_0 gives a negative current. Since this is impossible with a positive voltage source and, furthermore, is inconsistent with our assumption of a forward biased diode, this assumption must be incorrect.

3.2.2 Simple diode applications

To show how versatile the diode is, we give here several simple applications. The first, shown in Fig. 3.20, is a *voltage dropper*. Before discussing the circuit, we note the introduction of a new circuit symbol, the *common* or *ground* symbol. This is shown connected to the right end of the resistor. Ground serves as a common reference point for all other voltages referred to on the circuit diagram (recall that voltage is always between *two* points). Thus, for example, V_0 in this circuit is the voltage relative to ground. This might be provided by a battery of voltage V_0 connected between ground and the point labeled V_0.

Now let's get back the the voltage dropper. This might be used in a circuit where you have available one voltage (say, a 9 V battery) and need a slightly lower voltage. Each diode in the chain drops the voltage by approximately 0.6 V as shown. Note that this circuit makes no sense if we use the approximation $V_d = 0$.

The next example is the *limiter* (or *clipper*) circuit shown in Fig. 3.21 which is used to insure that the output voltage never exceeds a certain level, thus protecting the circuitry that follows from high voltage spikes or fluctuations. The diodes only come into play if they are forward biased; if a diode is not forward biased, no current flows through it and the voltage across it is unrestricted, just as if the

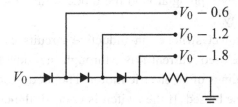

Figure 3.20 Voltage dropper.

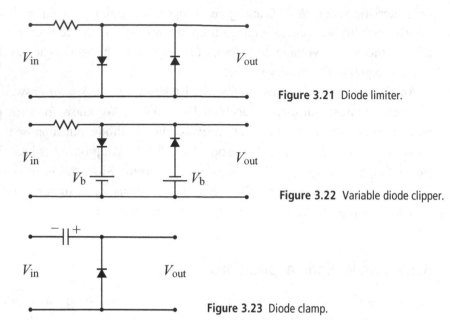

Figure 3.21 Diode limiter.

Figure 3.22 Variable diode clipper.

Figure 3.23 Diode clamp.

diode were not present. In order to forward bias a diode we must have the input voltage magnitude greater than 0.6 V. When this happens one of the diodes begins to conduct (depending on the polarity of the input voltage) and holds the voltage across itself to 0.6 V. Thus the output voltage (which is taken across the diode) can never exceed ± 0.6 V. A variable limiting level can be achieved by adding a battery to the circuit as shown in Fig. 3.22. Now when a diode conducts the output is held to a level $V_b + 0.6$.

The *clamp circuit* shown in Fig. 3.23 is used to shift an AC signal by a constant voltage. If the input voltage is less than -0.6 V, the diode can conduct and charge up the capacitor to a voltage $V_p - 0.6$, where V_p is the peak value of the AC voltage. Once this happens, the capacitor cannot discharge because to do so current would have to flow through the diode in the wrong direction. From this point on, therefore, the capacitor has a constant voltage across it with the polarity shown, so $V_{out} = V_{in} + V_c$. The output voltage is thus shifted up by a constant amount. An example case for a sinusoidal input is shown in Fig. 3.24. Finally, note that the shift can be made negative by flipping around the diode so that the capacitor charges with the opposite polarity.

Diodes are often used to protect switches in inductive circuits as shown in Fig. 3.25. When the switch is closed, current flows through the inductor, which might be, for example, the windings of an electric motor. The diode does nothing at this point because it is reverse biased. If the switch is opened, the current that

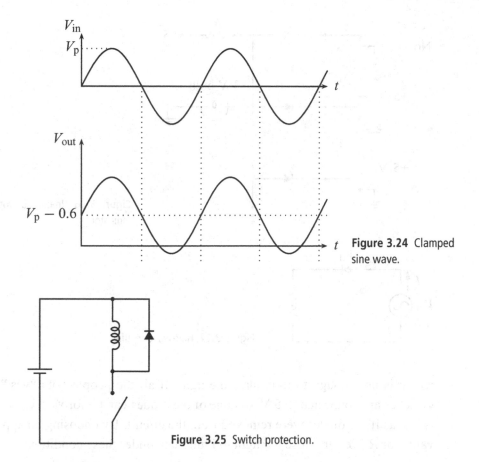

Figure 3.24 Clamped sine wave.

Figure 3.25 Switch protection.

was flowing must come to an abrupt halt. Since the voltage across the inductor is $L\frac{dI}{dt}$, the abrupt change in current produces a very large voltage, often large enough to produce an arc across the opening switch. If the switch is used frequently, this arcing will damage the switch. To prevent this, a diode is placed across the inductor. The polarity of the large induced voltage is such as to forward bias the diode and cause it to conduct, thus shorting out the inductor and protecting the switch.

Diodes can also be used to make logic circuits. An example of such usage is shown in Fig. 3.26 where we employ AND logic to form a unanimous vote indicator.[3] The circuit has a number of switches (any number is possible) that are connected to the 5 V supply when a person votes "yes" and connected to ground when a person votes "no." The indicator light requires 3 V to illuminate. If *any* switch is connected to ground, a circuit is completed which forward biases the diode connected to that switch. The voltage drop across that diode is then 0.6 V,

---

[3] We will have much more to say about logic circuits when we study digital electronics in Chapter 8.

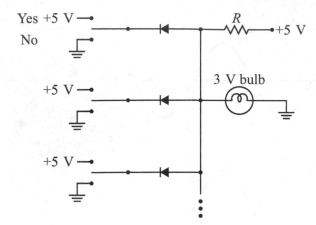

Figure 3.26 Unanimous vote indicator.

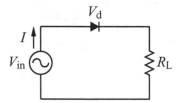

Figure 3.27 Half-wave rectifier.

which is not enough to illuminate the light. If all the people vote "yes," all the switches are connected to 5 V, so none of the diodes can be forward biased, and it is just as if the diodes were removed from the circuit. By choosing an appropriate value for R, we can cause the light to illuminate under these conditions.

3.2.3 Rectification

A major use for the diode is *rectification*, or making an alternating signal unidirectional. This is the first step in creating a DC power supply and is also used in AM radio receivers and other circuits.

The simplest rectifier, the *half-wave rectifier*, is shown in Fig. 3.27. If we are making a power supply, the AC signal source in this figure would represent the secondary of a transformer which takes the AC voltage coming out of the wall socket and changes it to another value suitable for our purpose. Except for the voltage source, this circuit is identical to that considered before (cf. Fig. 3.16), so the analysis used for that circuit can be employed here. Recall that the current through the diode will be zero until the voltage source exceeds 0.6 V, after which it will be

$$I = \frac{V_0 - 0.6}{R_L}. \tag{3.15}$$

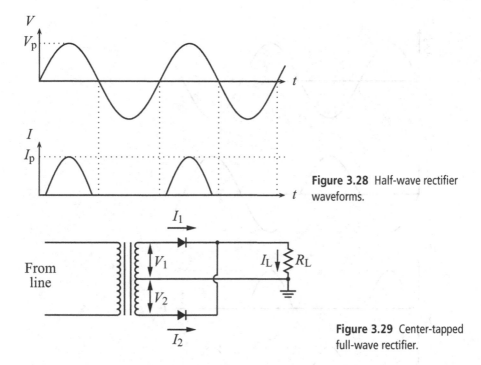

Figure 3.28 Half-wave rectifier waveforms.

Figure 3.29 Center-tapped full-wave rectifier.

For a voltage source $V_{in} = V_p \sin \omega t$, then, we expect a current waveform like that in Fig. 3.28, with peak value given approximately by $I_p = (V_p - 0.6)/R_L$. Note that the current flows only in one direction, as opposed to the alternating direction of the current if the diode were not present. As a consequence, the current (and, thus, the voltage across the load resistor) now has a non-zero average. If we think of the goal of creating a constant voltage supply, this is a step in the right direction. Of course, our waveform is still very bumpy and far from the constant voltage desired of a power supply, but we will see later how to fix this.

Our half-wave rectifier is simple (only uses one diode), but not very efficient. We essentially throw away half of our AC voltage and use only the part that forward biases our diode. The resulting waveform is also very bumpy, and this will make it more difficult to smooth out. An attempt to address these problems is shown in the *center-tapped full-wave rectifier* circuit of Fig. 3.29. This circuit uses two diodes and a specially altered transformer. The transformer has an additional wire attached at the middle of the secondary windings. This is called the *center tap* and gives the transformer user the option of using two identical sets of secondary windings.

The resulting waveforms for our center-tapped full-wave rectifier are shown in Fig. 3.30. We show the voltage for each half of the secondary relative to the grounded center tap. Note that these two voltages are 180 degrees out of phase; the voltage induced in each is the same, but we have tied different ends of the two

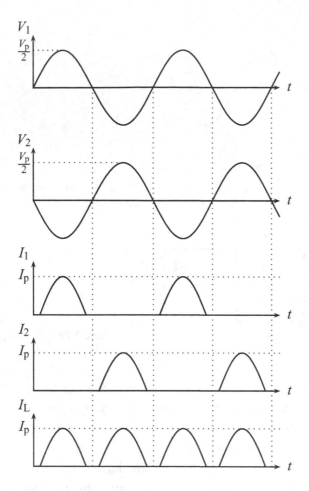

Figure 3.30 Center-tapped full-wave rectifier waveforms.

secondaries to ground, thus creating the inversion. Also, to have a fair comparison with our half-wave circuit, we assume the signal from each half of the secondary has a peak amplitude of $V_p/2$ since it uses only half of the total secondary windings. The circuit can then be viewed as essentially two half-wave rectifiers. When V_1 is positive, a current I_1 flows through the top diode, through the load resistor R_L and back through the center tap. On alternate half-cycles of the AC signal, a current I_2 flows through the bottom diode, through the load resistor R_L and back through the center tap. The key feature is that the load resistor now has current flowing through it on both half-cycles of the AC signal, and that current always flows in the same direction. There has been a cost, however, for obtaining a less bumpy current through the load resistor: the peak value of the current is now $I_p = \left(\frac{1}{2}V_p - 0.6\right)/R_L$ since we are using only half of the transformer at a time. We have thus not really fixed the efficiency problem.

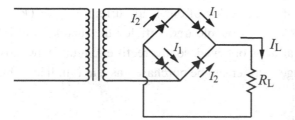

Figure 3.31 Full-wave bridge rectifier.

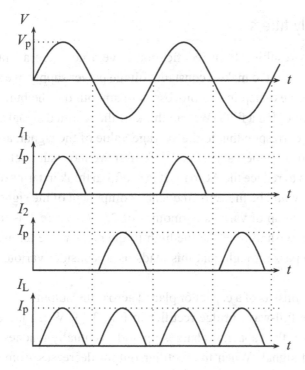

Figure 3.32 Full-wave bridge rectifier waveforms.

Our final rectifier circuit uses four diodes in a *full-wave bridge rectifier* configuration, as shown in Fig. 3.31. No center tapping of the secondary is necessary, so the voltage source again has peak value V_p as shown in the waveforms of Fig. 3.32. When the voltage source is positive (and exceeds the 1.2 V needed to turn on two diodes), current I_1 goes through the upper right diode, down through the load resistor, and back to the transformer through the lower left diode. When the voltage source is negative (and, thus, the bottom of the secondary is positive relative to the top), current flows through the bottom right diode, down through the load resistor, and back to the top of the secondary through the upper left diode. Now current flows through the diode on each half-cycle and the full secondary of the transformer is used. The only downside is that we need four diodes (not a problem since silicon diodes are cheap) and that we lose 1.2 V due to the voltage drop across two diodes rather than one. Thus the peak value of our load current is $I_p = (V_p - 1.2)/R_L$. For most circuits, this is a reasonable trade-off for the increased efficiency and reduced load current variation obtained, so this rectifier circuit is the most popular, and the full-wave bridge rectifier can be purchased as a unit in lieu of buying four individual diodes.

3.2.4 Power supply filters

While the waveforms resulting from rectification have a non-zero average, they still vary in time. If we want to make a constant voltage power supply, we still have some work to do. The next step in this process is to smooth out the bumps in our rectified signal by employing a filter. We can think of the rectified signal as having two parts: a DC part, corresponding to the average value of the signal, and an AC part, corresponding to the time-varying part left after we have subtracted the DC part. Our goal, then, is to reduce the AC part. It is useful to think in terms of Fourier analysis: the DC part would be the zero-frequency component of the signal and the AC part would be made up of various harmonics of $2\pi/T$, where T is the period of the signal. We then wish to filter out the high frequencies while letting the low frequency (i.e., zero) pass through, and this leads us to consider various types of low-pass filters.

The simplest filter consists of a capacitor placed across the output of the rectifier. This is shown for the full-wave bridge rectifier in Fig. 3.33. When the output of the rectifier is positive, the capacitor charges up and eventually reaches the peak value of the rectified signal. When the rectifier output decreases from its peak, the capacitor cannot follow because this would involve a discharge through the diodes; this cannot happen because of the orientation of the diodes. The capacitor *can* discharge, however, through the load resistor R_L, and the time scale for this process is set by the time constant $R_L C$. If we make this time constant large compared to the period of the rectified signal, there will be little change in the capacitor voltage before it is charged up again by the next cycle of the rectifier

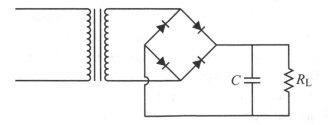

Figure 3.33 Simple capacitor filter.

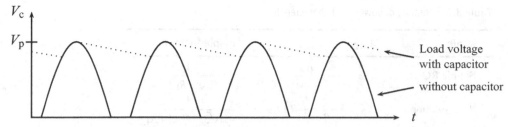

Figure 3.34 Simple capacitor filter waveforms.

output. This is shown in Fig. 3.34. Note that the voltage across the capacitor (or across the load) is now less bumpy and more like a constant.

When evaluating the quality of a power filter circuit, it is common to examine the *ripple factor*, r, which is defined as the ratio of the rms value of the AC component of the filtered signal to the DC or average value of the signal:

$$r = \frac{V_{\mathrm{rms,AC}}}{V_{\mathrm{DC}}}. \tag{3.16}$$

For the simple capacitor filter, it can be shown that, for $R_L C \gg T$, the ripple factor is given by

$$r \approx \frac{T}{2\sqrt{3}R_L C} = \frac{1}{2\sqrt{3}fR_L C} \tag{3.17}$$

where f is the frequency of the rectified signal. Examining this result, we see that r decreases with increasing C, R_L, and f. This is reasonable since increasing any of these parameters will decrease the amount the capacitor discharges. It can also be shown that, under these conditions, the DC part of the rectified signal is given by

$$V_{\mathrm{DC}} \approx V_p - \frac{V_p T}{2R_L C} \approx V_p - \frac{I_{\mathrm{DC}}}{2fC} \tag{3.18}$$

where V_p is the peak value of the rectified signal and, in the last equality, we have used the approximation $I_{\mathrm{DC}} \approx V_p/R_L$.

This last result draws attention to another factor to consider when judging the quality of a voltage source, the *regulation*. An ideal voltage source will supply a constant voltage regardless of the current supplied. As the last part of Eq. (3.18) shows, this is not the case for the simple capacitor filter: here V_{DC} decreases linearly with I_{DC}, although this non-ideal behavior improves with increasing f and C. We say that this power supply is *poorly regulated* because the output voltage depends on the output current.

As we will see in the next sections, neither ripple nor poor regulation is a huge problem these days because of the easy availability of solid state regulators.

Table 3.1 Summary of power filter characteristics

Name	V_{DC}	Ripple factor r
Simple RC	$V_p - \dfrac{I_{DC}}{2fC}$	$\dfrac{1}{2\sqrt{3}fR_L C}$
RC π-section	$V_p - \left(\dfrac{1}{2fC_1} + R\right)I_{DC}$	$\dfrac{1}{4\pi\sqrt{3}f^2 C_1 C_2 RR_L}$
LC or L-section	$\dfrac{2}{\pi}V_p$	$\dfrac{\sqrt{2}}{3\omega^2 LC}$

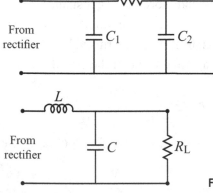

Figure 3.35 The RC π-section filter.

Figure 3.36 The LC or L-section filter.

Before we move on, however, it is worth giving two other examples of power supply filters, each having some desirable features. These are shown in Figs. 3.35 and 3.36 and expressions for the DC voltage and ripple factor are given in Table 3.1 (these approximate expressions are all derived under the assumption that the ripple is small, which is usually the desired condition). Note that both the π-section and L-section filters have ripple that goes like $1/f^2$, so these configurations are advantageous when the rectified signal has a high frequency (as might be the case with, for example, a gasoline powered generator). Also note that the L-section filter has good regulation: the DC voltage does not depend on the DC current at all.

3.2.5 Zener diodes

Although the L-section filter has perfect regulation in theory (i.e., V_{DC} has no dependence on I_{DC}), in reality the inductor will have some resistance and this will cause the regulation to degrade. In practice, one uses special circuits on the output of the power filter to provide improved regulation. There are various circuits that can be employed in this effort, but the simplest involves a special kind of diode called a *zener diode*.

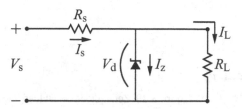

Anode ——————▶⊢—————— Cathode   **Figure 3.37** Electronic symbol for a zener diode.

Figure 3.38 Zener diode circuit.

Recall that a diode with sufficient reverse bias will undergo breakdown. When this happens, the reverse current through the diode rises rapidly while the voltage across the diode remains roughly constant. This is reminiscent of our goal for a well regulated power supply: a constant voltage that does not vary with current. In fact, this property is so useful that special diodes are manufactured that have extremely sharp I–V curves at breakdown. The symbol for this *zener diode* is shown in Fig. 3.37. Such diodes are specified by their breakdown voltage V_b and their maximum power rating $P_{max} = V_b I_{max}$.

Consider the typical zener diode circuit shown in Fig. 3.38. The voltage V_s and resistance R_s can be thought of as representing the Thevenin equivalent for the output of the filter circuit. To this we attach the parallel combination of a zener diode and a load resistor. Note that the zener is installed with an orientation that looks backwards: if this were a normal diode, it would never conduct since the polarity of V_s will not allow current to flow in the direction of the diode triangle. In this case, however, we are interested in operating the diode in its breakdown mode, so the reverse orientation is appropriate.

Applying our usual circuit laws to this circuit, we obtain $V_s - I_s R_s - V_d = 0$, $I_s = I_z + I_L$, and $V_d = I_L R_L$, where V_d is the voltage across the diode and I_z is the current through the diode. Combining these we obtain

$$\frac{V_s - V_d}{R_s} = I_s = I_z + \frac{V_d}{R_L} \tag{3.19}$$

which, after some manipulation, can be cast in the form

$$\frac{V_s}{R_s} - \left(\frac{R_L + R_s}{R_s R_L}\right) V_d = I_z. \tag{3.20}$$

Since I_z is a complicated function of V_d, we again turn to a graphical load line analysis. The straight line resulting from Eq. (3.20) is plotted along with the diode I–V characteristic in Fig. 3.39. Because we normally intend to use the zener diode in reversed orientation, we have inverted the normal diode I–V characteristic.

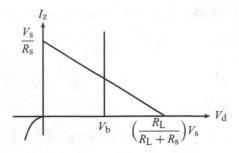

Figure 3.39 Zener diode circuit load line analysis.

Referring to Figs. 3.38 and 3.39, we can draw a number of conclusions.

1. In order for the zener to operate in breakdown mode (i.e., for the operating point to be located on the breakdown portion of the I–V characteristic), we must have

$$\left(\frac{R_L}{R_s + R_L}\right) V_s > V_b \tag{3.21}$$

where V_b is the breakdown voltage.

2. The following points apply while the zener is in breakdown mode.
   (a) As always, $I_s = I_z + I_L$. When in breakdown mode, $I_s = (V_s - V_b)/R_s$, so if V_s is fixed, so is I_s. Thus if we vary the load resistance (and thus I_L), I_z must change in the opposite way. If I_L increases, I_z will decrease, and vice versa.
   (b) In order to handle any possible case (including $R_L \to \infty$), the zener diode should be rated high enough to handle *all* of I_s.
   (c) If the voltage V_s changes, the voltage across the load remains constant at V_b. This is just what we want for our DC power supply. The load current $I_L = V_b/R_L$ remains fixed, too. Any changes in V_s are taken up by changes in I_s and I_z.

3. If the zener is reverse biased but not in breakdown mode (i.e., if Eq. (3.21) is not satisfied), the operating point will be $I_z \approx 0$ and

$$V_d = \left(\frac{R_L}{R_s + R_L}\right) V_s. \tag{3.22}$$

This is just the voltage divider equation. Hence in this case the circuit operates as if the zener were not present.

4. If the zener is forward biased, it acts like a regular diode.

5. While the breakdown portion of the I–V characteristic is drawn in Fig. 3.39 as a vertical line (i.e., with infinite slope), it actually has a finite slope. This is specified by quoting the inverse of the slope, $\Delta V_d/\Delta I_z$. This is called the *dynamic resistance* and is typically about 1 Ω for a zener diode.

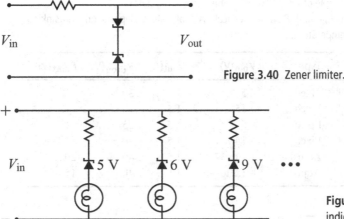

Figure 3.40 Zener limiter.

Figure 3.41 Zener DC voltage indicator.

The zener diode has other applications beyond its use as a regulator. Two examples are given in Figs. 3.40 and 3.41. The zener limiter circuit, like the limiter and clipper circuits we have seen before, places limits on the output voltage. Two zener diodes are connected in series across the output with opposite orientations as shown in Fig. 3.40. When the input voltage is positive and exceeds $V_b + 0.6$ V, the diodes conduct, with the top diode forward biased (and thus with a voltage drop of 0.6 V) and the bottom diode reverse biased and in breakdown mode. If the input voltage increases, the output is held at $V_b + 0.6$ V. Similarly, if the input is more negative than $-(V_b + 0.6$ V$)$, the bottom diode is forward biased with 0.6 V drop and the top diode is reverse biased with a voltage drop of V_b. Thus the output can never be less than $-(V_b + 0.6$ V$)$ or greater than $V_b + 0.6$ V.

Figure 3.41 shows how zeners can be used to make a simple voltage indicator. The circuit makes use of the fact that zener diodes can be obtained with a variety of breakdown voltages, and the diodes in the circuit are labeled with their respective breakdown voltages. As the input voltage increases, we eventually reach the point where the leftmost zener can break down, and the resulting current flow turns on the first light. At a higher voltage, the next zener also breaks down and its light also turns on, and so on. Thus for higher input voltage, more of the indicator lights are on. This type of display is often seen on stereo systems and cell phones to give, for example, an indication of the received signal strength.

3.2.6 Regulators

While a zener diode can be employed to make a simple and inexpensive regulator, its non-zero dynamic resistance means that the output voltage will still vary slightly

Table 3.2 A sample of fixed and adjustable voltage regulators
The values for I_{out} and P_{max} are typical. Actual values depend on heatsinking and ambient temperature.

Number	Type	V_{out}(V)	I_{out}(A)	V_{drop}(V)	P_{max}(W)
LM7805CT	Fixed	+5	1.5	30	1.7
LM7815CT	Fixed	+15	1.5	20	1.7
LM7905CT	Fixed	−5	1.5	25	1.7
LM7915CT	Fixed	−15	1.5	30	1.7
LM317T	Adjustable	1.2 to 37	1.5	40	20
LM337T	Adjustable	−1.2 to −37	1.5	40	15

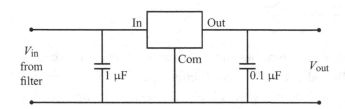

Figure 3.42 Usage example for a fixed voltage regulator.

with output current. For more demanding applications, a more sophisticated voltage regulator is employed. While we are not ready to understand the inner working of these devices, we can understand how to use them to finish our DC power supply design. A very small sample of typical voltage regulators is given in Table 3.2.

Regulators are specified by their output voltage (or, in the case of variable types, output voltage range) and their power rating $P_{max} = V_{drop}I_{out}$, where V_{drop} is the maximum voltage difference between the output and the input, and I_{out} is the output current. They usually come in a package with three leads, and a typical circuit for a fixed output regulator is shown in Fig. 3.42. The input voltage would typically be the output of the power filter. In order for the regulator to work properly, this input voltage must at all times be a certain level (typically 3 V) above the level of the output voltage.[4] On the other hand, you do not want the input voltage to be too high above the output because this would give a large V_{drop} and thus necessitate a large power rating (and increased expense). The capacitors on the input and output of the voltage regulator are specified by the manufacturer to eliminate high frequency noise and provide stability for the regulator.

Regulators with variable output voltage are also available, and a typical circuit for this type is shown in Fig. 3.43. The variable resistor R_2 provides the adjustment and the resulting output is given by $V_{out} = 1.25(1 + R_2/R_1)$. Rules similar to those

---

[4] This assumes a positive output. For negative output, the input must be 3 V below the output level.

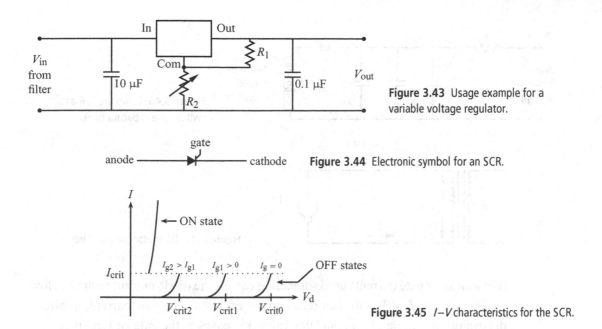

Figure 3.43 Usage example for a variable voltage regulator.

Figure 3.44 Electronic symbol for an SCR.

Figure 3.45 I–V characteristics for the SCR.

for the fixed regulator apply here, so the maximum output voltage will be about 3 V below the minimum input voltage. Because the output is now variable, V_{drop} will also be variable, so one must be careful to choose a power rating that is sufficient for the worst case.

3.2.7 The silicon controlled rectifier

The silicon controlled rectifier or SCR is, as the name implies, a diode that can be, to some extent, controlled. The schematic symbol for this device is shown in Fig. 3.44. The control is provided by a third lead called the *gate*.

As with most electronic devices, the behavior of the SCR can be seen in its I–V characteristic, but now, since there is a third independent lead, there are an infinite number of I–V characteristics, one for each value of the gate parameter. In such cases, it is customary to display a family of curves for representative values of the control parameter on the same plot. This is shown for the SCR in Fig. 3.45. The relevant control parameter is the *current I_g* flowing into the gate.

The SCR can be thought of as having an "on" state and an "off" state. The device is normally off. As the voltage across the diode V_d is increased, the diode stays off (allowing no current flow) until a critical voltage V_{crit} is approached. The current then increases slightly until the critical voltage is reached, at which point the diode switches to its on state. Now it operates like a normal diode, with a large forward current and a small forward voltage drop. The SCR will remain in this on state

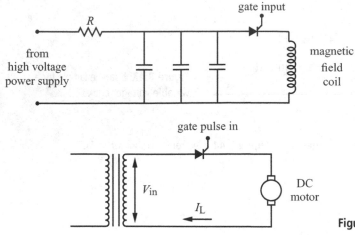

Figure 3.46 Using the SCR as a switch for a capacitor bank.

Figure 3.47 DC motor speed control.

(regardless of gate current) until something causes the diode current to drop below a critical level, I_{crit}. It will then return to the off state. The gate current, in effect, determines the value of V_{crit}, and this value decreases as the gate current increases, as shown in Fig. 3.45. The value of the gate current necessary to turn on the SCR is typically much smaller than the maximum current allowed through the on-state diode, so the SCR allows the user to control a high power circuit with a low power circuit.

The SCR can be used in a variety of ways. Figure 3.46 shows its use as an electrical switch in a pulsed high magnetic field system. A bank of capacitors is charged to a high voltage.[5] The SCR in the circuit is chosen so that the high voltage is less than V_{crit0}. This prevents current from flowing through the field coils while the gate current is zero. When the user is ready, a current pulse is applied to the gate that is sufficient to lower the critical voltage below the voltage across the SCR. The SCR thus switches to its on-state and the capacitors discharge through the magnetic field coil, producing the desired high magnetic field. If the resistor R is large enough, the current through the SCR will eventually fall below I_{crit} and the SCR will switch off.

A second example of SCR usage is shown in Fig. 3.47 where we show a simple DC motor speed control. The speed of a DC motor depends on the average of the current in the circuit. Here a pulse is applied to the gate at an adjustable time. When the pulse is applied, the SCR turns on and rectifies the sine wave input voltage. The earlier the pulse is applied, the larger the average current, and the higher the speed of the motor. When the rectified current goes to zero, the SCR turns off and

---

[5] A high voltage is used to maximize the stored energy since $E = \frac{1}{2}CV^2$.

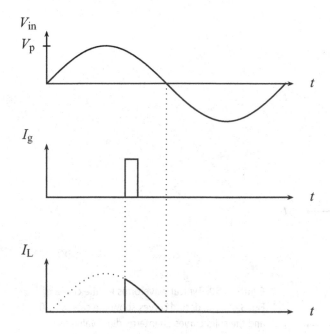

Figure 3.48 DC motor speed control waveforms. For comparison, the dotted line for I_L shows the waveform for an ordinary diode.

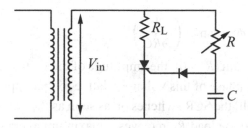

Figure 3.49 Control of the SCR switch time with an RC circuit.

awaits the next gate pulse. Some representative waveforms are shown in Fig. 3.48. This circuit can be used to control anything that depends on the average current in the circuit.

The DC motor control requires additional circuitry to produce the required gate pulse. The next example (see Fig. 3.49) shows how the SCR can be controlled without elaborate external circuitry. An adjustable RC circuit is used to control the amplitude and the phase of the voltage applied across the gate-cathode junction of the SCR. The diode attached to the gate protects the SCR gate from reverse bias.

From our previous analysis of the RC circuit driven by a sine wave of the form $V_p \sin \omega t$, we know that the voltage across the capacitor is

$$V_c = \frac{V_p}{\sqrt{1 + (\omega RC)^2}} \sin\left(\omega t + \phi - \frac{\pi}{2}\right) \tag{3.23}$$

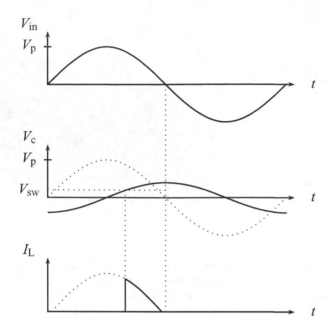

Figure 3.50 Typical waveforms for the circuit of Fig. 3.49. The dotted curves show the case $R \approx 0$ and the solid curves an intermediate value of R.

where

$$\phi = \tan^{-1}\left(\frac{1}{\omega RC}\right). \tag{3.24}$$

When R is adjusted to its minimum ($R \approx 0$), the amplitude of the voltage across the capacitor is maximum and the phase of this voltage relative to that applied across the SCR is near zero. As a result, the SCR switches on as soon as the voltage across the SCR becomes positive, and the load R_L receives a maximum average current. This is shown in Fig. 3.50, where we assume a gate voltage V_{sw} is necessary to turn the SCR on. When R is adjusted towards its maximum, the amplitude of the voltage across the capacitor is reduced and $\phi \approx 0$ so the phase of this voltage is shifted by $-\frac{\pi}{2}$. This combination of reduced amplitude and increased phase shift delays the SCR firing time by half a period, and by this time the voltage across the SCR has become negative so the SCR never turns on. Intermediate values of R produce intermediate turn-on times and one of these is shown in Fig. 3.50. Thus by adjusting R one can vary the average current through the load from its maximum to zero.

EXERCISES

1. Sketch the energy band configuration for a p-n juntion under the following conditions: no bias, reverse bias, and forward bias. On each, indicate the direction of the net current.
2. Sketch the output waveforms expected when a 100 Hz, 5 V_p sine wave is applied to each of the circuits in Fig. 3.51. Specify important voltage levels and time scales. The input is on the left and the output is on the right.
3. Sketch the output waveforms expected when a 1000 Hz, 20 V_{pp} square wave is applied to each of the circuits shown in Fig. 3.51. Specify important voltage levels and time scales. The input is on the left and the output is on the right.

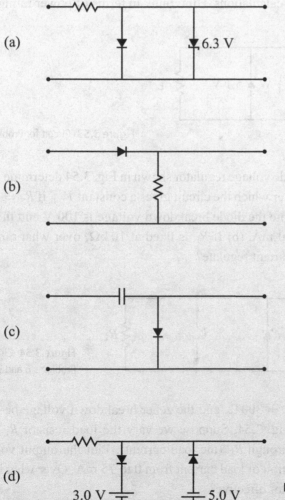

(a)

(b)

(c)

(d)

3.0 V   5.0 V

Figure 3.51 Circuits for Problems 2 and 3.

4. Determine the necessary values of the components (R, C_1, and C_2, and V_p for the transformer output) for the circuit of Fig. 3.52, with the requirements that the load voltage be 15 V with 0.01% or less ripple and the load current be 100 mA.

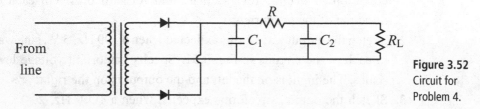

From line

Figure 3.52
Circuit for
Problem 4.

5. (a) In the circuit of Fig. 3.53, what would happen if the load resistor were shorted? (b) What would happen if the load resistor were removed? Support your answers with calculations. Hint: think in terms of power ratings.

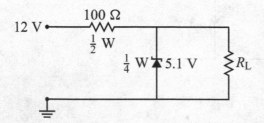

Figure 3.53 Circuit for Problem 5.

6. (a) In the zener diode voltage regulator shown in Fig. 3.54 determine the range of load resistances over which the circuit gives a constant V_{out} if $R_s = 1500\ \Omega$ and $V_s = 150$ V. Assume the diode breakdown voltage is 100 V and the maximum rated current is 100 mA. (b) If R_L is fixed at $10\ k\Omega$, over what range of input voltages does the circuit regulate?

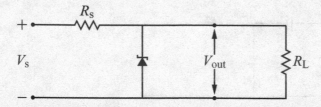

Figure 3.54 Circuit for
Problems 6 and 7.

7. Let $V_s = 30$ V, $R_s = 300\ \Omega$, and the zener breakdown voltage be 15 V in the circuit shown in Fig. 3.54. Suppose we vary the load resistor R_L in order to vary the current through R_L (the load current). Plot the output voltage of the regulator as a function of load current from 0 to 75 mA. Over what load current range is the regulator effective?

8. Design a variable voltage power supply capable of supplying 250 mA with output voltage range 5 to 15 V. Specify the relevant details (e.g., values, model numbers, power rating) for all the components you use.

FURTHER READING

James J. Brophy, *Basic Electronics for Scientists*, 5th edition (New York: McGraw-Hill, 1990).

Charles Kittel, *Introduction to Solid State Physics*, 4th edition (New York: Wiley, 1971).

John E. Uffenbeck, *Introduction to Electronics, Devices and Circuits* (Englewood Cliffs, NJ: Prentice-Hall, 1982).

M. Russell Wehr, James A. Richards, Jr., and Thomas W. Adair, III, *Physics of the Atom*, 4th edition (Reading, MA: Addison-Wesley, 1985).

4 Bipolar junction transistors

4.1 Introduction

The silicon controlled rectifier introduced in the last chapter was the first device we have seen that offered some measure of electronic control: the gate current determined the details of the I–V characteristic. This control, however, was fairly limited. In the examples we considered, the gate current determined the time at which the SCR switched to its on-state. Once the SCR was turned on, however, its behavior was no longer related to the magnitude of the gate current, and removing the gate current altogether would not return the SCR to its off-state.

We now turn to a device with a greater measure of electronic control: the *transistor*. Like the SCR, the transistor allows the user to control a large current through the device with a smaller control signal. But with the transistor, one can have proportional control; that is, the amount of current through the device is proportional to the control signal. This allows one to *amplify* signals, which is fundamental to all types of electronic communication.

Transistors come in two basic types: bipolar junction transistors (BJTs) and field-effect transistors (FETs). This chapter will cover the fundamentals of BJTs and also introduce some common terminology for transistor amplifiers. FETs are addressed in Chapter 5.

4.2 Bipolar transistor fundamentals

A bipolar transistor can be thought of as a sandwich of n-type and p-type semiconductors. Of course, there are two ways to form this sandwich: a piece of p-material between two pieces of n-material (called an *npn transistor*), or a piece of n-material between two pieces of p-material (called a *pnp transistor*). The circuit symbols for these transistors are shown in Fig. 4.1. We will focus here on the npn transistor; the development for a pnp transistor is similar except the polarities and current directions are reversed.

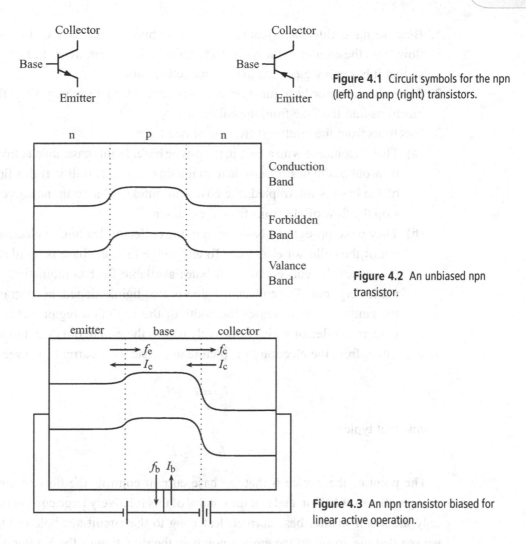

Figure 4.1 Circuit symbols for the npn (left) and pnp (right) transistors.

Figure 4.2 An unbiased npn transistor.

Figure 4.3 An npn transistor biased for linear active operation.

The band structure for our npn sandwich is shown in Fig. 4.2. The same shift in energy levels charateristic of the p-n diode is seen here. Although the transistor looks symmetric, it is not: as we will see below, the two n-materials are doped differently.

We now apply external voltages between the junctions. Although there are various ways to do this, a useful and common situation is shown in Fig. 4.3. As indicated, the leftmost p-n junction is forward biased and the rightmost junction is strongly reverse biased. The three parts of the transistor are labeled as shown: emitter, base, and collector.

The operation of the transistor when biased this way can be described as follows.

1. Because the emitter-base junction is forward biased, electrons will have a net flow from the emitter to the base. To facilitate this, the emitter is heavily doped so that it has many electrons in the conduction band.
2. Since the collector-base junction is reverse biased, there is very little flow of electrons into the base from the collector.
3. Electrons from the emitter suffer one of two fates.
   (a) They recombine with a hole in the p-type base. In this case, an electron must flow out of the base electrode to maintain charge neutrality. If this flow out of the base is interrupted, the base will quickly charge up negatively and stop the flow of electrons from the emitter.
   (b) They pass through the base and into the collector, leading to electron flow out of the collector electrode. To encourage this, the base is lightly doped (to reduce the concentration of holes available for recombination) and is made very thin. The collector region is also lightly doped, in contrast with the emitter. This increases the width of the depletion region between the base and collector which effectively makes the p-material even thinner.
4. Switching from the electron flow picture to considering currents, we see that

$$I_e = I_c + I_b \tag{4.1}$$

and that typically

$$I_b \ll I_e, I_c. \tag{4.2}$$

The point of the device is that the base current controls the flow of electrons from emitter to collector, and that the control of this relatively large current requires only the much smaller base current. Referring to the circuit symbols in Fig. 4.1, we see that the arrow on the emitter points in the direction of the emitter current; for the npn transistor, the emitter current is out of the transistor. The base and collector currents have the opposite orientation to the emitter current; thus, for the npn transistor, these currents go into the transistor. The opposite current directions apply to the pnp transistor.

Two dimensionless parameters that characterize the relationships between these currents are α and β. The parameter α is the fraction of the emitter current that makes it to the collector, thus

$$I_c = \alpha I_e \tag{4.3}$$

and, combining this with Eq. (4.1),

$$I_b = (1 - \alpha)I_e. \tag{4.4}$$

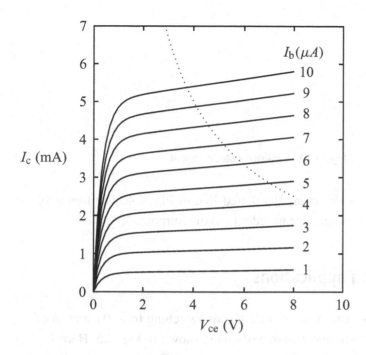

Figure 4.4 Typical *I–V* characteristic for a bipolar transistor showing the collector current I_c versus collector-emitter voltage V_{ce} with base current I_b as a parameter. The dotted line shows the power limit curve.

Note that $\alpha < 1$ necessarily, but typically α is close to one. The second way of parametrizing the transistor is with β, where

$$\beta \equiv \frac{I_c}{I_b} = \frac{\alpha}{1-\alpha}. \tag{4.5}$$

Typically, $\beta \gg 1$.

There are various $I–V$ curves one can make for the transistor, but one of the most useful is a plot of the collector current I_c versus the voltage between the collector and the emitter V_{ce}. A different curve is obtained for each value of the base current I_b, so in practice a family of $I–V$ curves is plotted as shown in Fig. 4.4. In the broad central portion of the plot, the collector current is roughly proportional to the base current. This is called the *linear active region* and corresponds to the biasing scheme shown in Fig. 4.3 and described by Eqs. (4.3), (4.4), and (4.5). On the left side of the plot all the different base current lines converge to roughly the same collector current, so proportionality no longer holds. This is the *saturation region* and occurs when the collector-base junction is no longer strongly reverse biased and some electrons can flow into the base from the collector, thus reducing I_c. As the base current approaches zero (because the emitter-base junction is no longer forward biased), the collector current also falls to zero. This *cutoff region* is at the bottom of the plot. Like most components, the transistor has a maximum power rating P which is equal to the maximum product of V_{ce} and I_c. A typical plot of

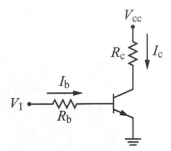

Figure 4.5 Transistor switching circuit.

$I_c^{\max} = P/V_{ce}$ versus V_{ce} is shown by the dotted line in Fig. 4.4. The transistor must be operated to the left of this line in order to avoid burnout.

4.3 DC and switching applications

One application of the transistor is to provide DC or switching (on-off) control of a current. A simple but illustrative example of this is shown in Fig. 4.5. Here V_{cc} is the name given to a constant power supply voltage and V_1 is a control voltage, which, as we shall see, controls the flow of current through resistor R_c. Both voltages are understood to be relative to ground.

There are two circuit loops we must analyze to understand this circuit. The first starts with voltage V_1 and continues through the resistor R_b and across the base-emitter junction giving $V_1 - I_b R_b - V_{be} = 0$, where V_{be} is the voltage from the base to the emitter. Solving for I_b gives

$$I_b = \frac{V_1 - V_{be}}{R_b}. \tag{4.6}$$

The problem here is that I_b is a complicated function of V_{be}, so again we have a transcendental equation. We will approximate the behavior of the base-emitter junction as equivalent to that of a diode junction. Now we are on familiar ground, and a graphical solution of Eq. (4.6) is shown in Fig. 4.6. As V_1 is increased, the base current increases. As with the diode, we will often find it useful to approximate $V_{be} \approx 0.6$ V when the junction is forward biased.

We now turn to the second circuit loop in Fig. 4.5. Starting with the power supply voltage, we obtain $V_{cc} - I_c R_c - V_{ce} = 0$, where V_{ce} is the voltage from the collector to the emitter of the transistor. Solving here for I_c gives

$$I_c = \frac{V_{cc} - V_{ce}}{R_c}. \tag{4.7}$$

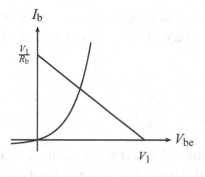

Figure 4.6 Graphical solution for the control loop of the transistor switching circuit.

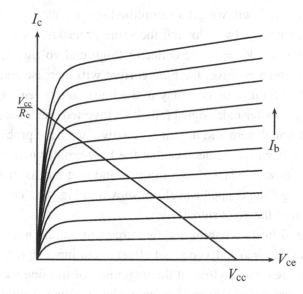

Figure 4.7 Graphical solution for the supply loop of the transistor switching circuit.

Again, we face a transcendental equation because I_c is a complicated function of V_{ce}. The graphical solution for this case is shown in Fig. 4.7. Analyzing Eq. (4.7) gives the y-intercept of the straight load line as V_{cc}/R_c and the x-intercept as V_{cc}. The solution of Eq. (4.7) is given by the intersection of the load line with the transistor characteristic curve. But which characteristic curve should we use? This depends on the value of the base current, and thus on the solution to the control circuit part of the problem.

Let's consider what will happen if we let V_1 switch between zero and some large, positive value. When V_1 is zero, our graphical analysis of Fig. 4.6 gives $I_b = 0$. Then Fig. 4.7 gives a solution $I_c = 0$ and $V_{ce} = V_{cc}$. When V_1 is positive, on the other hand, some base current will flow. If we choose V_1 and R_b so as to give I_b large enough, then the solution will lie on the left edge of Fig. 4.7, i.e., the transistor will be saturated. The current I_c is now large and V_{ce} is small. Our circuit thus gives us on-off control of the current through R_c by switching the control voltage V_1.

In addition, if we take the collector voltage as an output, we have a voltage inverter: a low input gives a high output, and vice versa. This is often useful in digital circuits.

4.4 Amplifiers

As noted above, an important application of transistors is the amplification of AC signals. If we apply a sinusoidal signal to the circuit of Fig. 4.5 and take the output from the collector to ground, will we get an amplified signal? We can answer this question, at least qualitatively, by following the same procedure as used above. We already know that when $V_1 = 0$, the collector-to-ground voltage V_{ce} will be V_{cc}. As V_1 starts to become positive, the base current will increase and V_{ce} will decrease. As long as we limit I_b so as to stay in the linear active region, V_{ce} will track V_1 and we will get a sinusoidal signal (albeit an inverted one – see Fig. 4.8). But when V_1 passes through zero and becomes negative, we have problems. The base current cannot be negative (remember that the base-emitter junction is like a diode) so it remains at zero when V_1 is negative, and that means that V_{ce} will remain at V_{cc}. We thus get the clipped waveform shown in Fig. 4.8. This is hardly a faithful amplification of the input signal.

The problem stems from the inability of the circuit to handle negative input signals. One way to fix this is to add a constant offset to our input signal, as shown in Fig. 4.9. We adjust the offset so that, at the beginning of the sine wave, V_{ce} is in the middle of the linear active region. Now as the input voltage swings through its cycle, the base current will move higher and then lower than its initial value but will never become negative (at least if we keep the input sine wave amplitude small enough). Thus V_{ce} will move lower and then higher than its initial value but

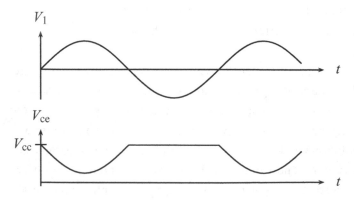

Figure 4.8 Clipped output results when we apply a sine wave to the switching circuit.

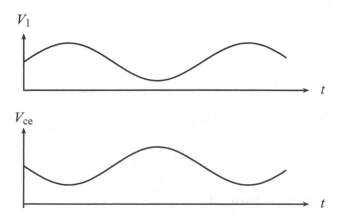

Figure 4.9 Adding a constant to the input signal removes the clipping.

will not clip as before, so our output is an inverted, offset sine wave. Although it is not evident from our discussion, the output is indeed amplified, as we shall see.

We now have a solution, but not a very practical one. Most communication signals, for example, do not come with an offset level but simply vary around zero. Rather than insist that our input signals have this offset, it makes more sense to change our circuit to handle signals that vary around zero, and this is our next task. We also need to decide what quantities are necessary to specify an amplifier's features and develop techniques for calculating these quantities.

4.4.1 The universal DC bias circuit

Our first task is to devise a circuit that will keep the transistor operating in the middle of the linear active region. This circuit is called the *universal DC bias circuit* because it will be used in all the amplifier circuits we consider and because it sets the constant or DC operating conditions. Later, we will add an AC signal to this circuit to produce an amplifier.

Our circuit is shown in Fig. 4.10. Power is supplied by the voltage source V_{cc}, while R_1 and R_2 form a voltage divider. The collector and emitter resistors, R_c and R_e, complete the circuit.

Another reason we call this circuit *universal* is that other bias circuits can be derived from this one by setting a resistor value either to zero (if it is replaced by a wire in the alternative circuit) or to infinity (if it is missing in the alternative circuit). The expressions we derive below can still be used with the appropriate adjustment of resistor values.

To simplify analysis, it is convenient to replace the left half of the circuit by its Thevenin equivalent, as indicated in Fig. 4.11. Our original circuit can then be

Bipolar junction transistors

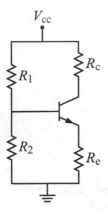

Figure 4.10 The universal DC bias circuit.

Figure 4.11 Thevenin equivalent for the left half of the universal DC bias circuit.

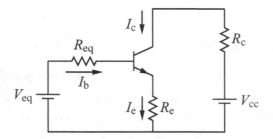

Figure 4.12 Redrawn version of the universal DC bias circuit.

redrawn as shown in Fig. 4.12. We would like to obtain equations for I_c, I_b, and V_{ce}; these three quantities define the *operating point* of the transistor (also called the *quiescent point* or *Q point*).

Applying KVL to the left side of Fig. 4.12 gives

$$V_{eq} - I_b R_{eq} - V_{be} - I_e R_e = 0 \tag{4.8}$$

while the right side gives

$$V_{cc} - I_c R_c - V_{ce} - I_e R_e = 0. \tag{4.9}$$

We also note that $I_e = I_c + I_b$ (which is always true) and

$$I_c = \beta I_b \tag{4.10}$$

(which is true for the linear active region where we wish to operate). Combining these gives $I_e = (\beta + 1)I_b$, and using this result in Eq. (4.8) yields

$$I_b = \frac{V_{eq} - V_{be}}{R_{eq} + (\beta + 1)R_e}. \tag{4.11}$$

Similarly, Eq. (4.9) produces

$$V_{ce} = V_{cc} - I_c \left(R_c + \frac{\beta + 1}{\beta} R_e \right). \tag{4.12}$$

Thus if we know the four resistor values of our circuit, β, the power supply voltage V_{cc}, and approximate V_{be}, we can determine the operating point from Eqs. (4.11), (4.12), and (4.10). We can also cast the equations in a form more convenient to use if we know the operating point we want and seek the appropriate resistor values:

$$I_b \left[\frac{R_1 R_2}{R_1 + R_2} + (\beta + 1)R_e \right] = \left(\frac{R_2}{R_1 + R_2} \right) V_{cc} - V_{be} \tag{4.13}$$

and

$$R_c + \left(\frac{\beta + 1}{\beta} \right) R_e = \frac{V_{cc} - V_{ce}}{I_c}. \tag{4.14}$$

Since we have two equations and four unknown resistances, there is no unique solution to our DC bias problem. As we will see later, other amplifier parameters will further restrain our choice of resistor values. For now, we note that a useful procedure is to start by choosing $R_1 = R_2 = 10\,k\Omega$ and then using Eq. (4.13) to solve for R_e and then Eq. (4.14) to solve for R_c.

4.4.2 Black box model of an amplifier

Before we get into the details of amplifier circuits and analysis, we should think about the things we would like to know about our amplifier. As we shall see, the simple amplifier model in Fig. 4.13 will allow us to use the amplifier effectively if we know the following quantities.

1. The open-loop voltage gain, a_{OL}. Generally, the voltage gain, a, is just V_{out}/V_{in}, but this will depend on the value of the load resistance and other things. We can obtain a parameter that just depends on the amplifier if we define

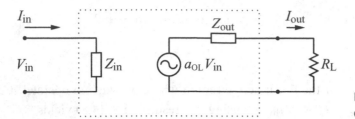

Figure 4.13 Black box model of an amplifier.

$$a_{OL} = \frac{V_{out}(R_L = \infty)}{V_{in}} \quad (4.15)$$

where $V_{out}(R_L = \infty)$ is the output voltage for the case where the load resistance is infinite. An infinite load resistance is the same as having nothing connected to the output terminals; hence, the term "open loop."

2. The current gain $g = I_{out}/I_{in}$. This parameter is less useful since I_{out} necessarily depends on the value of R_L.
3. The input impedance $Z_{in} = V_{in}/I_{in}$.
4. The output impedance Z_{out}:

$$Z_{out} = \frac{V_{out}(R_L = \infty)}{I_{out}(R_L = 0)} \quad (4.16)$$

where $I_{out}(R_L = 0)$ is the output current for the case where the load resistance is zero.

The reader can verify that these definitions are consistent with the model of Fig. 4.13. For example, if there is nothing connected to the output of the amplifier (which is equivalent to $R_L = \infty$), then no current will flow and there will be no voltage drop across Z_{out}. Thus the output voltage will be the same as the voltage source in the model, $a_{OL} V_{in}$. This is consistent with Eq. (4.15).

As an example of the usage of this *black box model*, consider the following problem. Suppose an amplifier has an open-loop voltage gain of 50, an input impedance of 100 Ω, and an output impedance of 10 Ω. The amplifier is driven with a sine-wave generator with output impedance 50 Ω and an open loop amplitude of 0.1 V_{rms}. Find the power gain in decibels when an 8 Ω load is attached to the amplifier output. Here the power gain is defined as the ratio of the power into the amplifier to the power into the load.

The first step in our solution is to draw the circuit corresponding to the situation described in the problem. This is shown in Fig. 4.14. Using Thevenin's theorem, the signal generator is modeled as an AC voltage source in series with a resistance. We are given the open loop amplitude of this generator, which means the voltage at

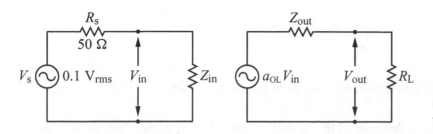

Figure 4.14 Example of using the black box model.

the generator output when nothing is attached; since no current flows in this case, there is no drop across the output impedance R_s, so this output voltage must be the same as the AC voltage source. We have also labeled the input and output voltages of the amplifier, V_{in} and V_{out}, respectively. The remaining features of our drawing follow from Fig. 4.13 and the problem description.

The power into the amplifier is just V_{in}^2/Z_{in}, where V_{in} is the rms amplitude. Since the input circuit forms a voltage divider, we have

$$V_{in} = \left(\frac{Z_{in}}{Z_{in} + R_s}\right) V_s = \left(\frac{100\ \Omega}{150\ \Omega}\right)(0.1\ V_{rms}) = 0.067\ V_{rms}. \tag{4.17}$$

Hence, the input power is $P_{in} = V_{in}^2/Z_{in} = 44.9\ \mu W$. Similarly, the output circuit gives

$$V_{out} = \left(\frac{R_L}{Z_{out} + R_L}\right) a_{OL} V_{in} = \left(\frac{8\ \Omega}{18\ \Omega}\right)(50 \times 0.067\ V_{rms}) = 1.49\ V_{rms} \tag{4.18}$$

giving an output power of $P_{out} = V_{out}^2/R_L = 0.28$ W. The power gain is thus $P_{out}/P_{in} = 6171$ or, in decibels, $10 \log (P_{out}/P_{in}) = 37.9$ dB.

4.4.3 AC equivalents for bipolar junction transistors

Having seen the usefulness of our black box model for an amplifier, our task is now to formulate a way to calculate the black box parameters a_{OL}, g, Z_{in}, and Z_{out}. Before we get into the details, let's review our approach. We know that, in order to avoid clipping, we need to operate the transistor in the middle of the linear active region, and we have devised the universal DC bias circuit to accomplish this. We now wish to add an AC signal to this circuit which the transistor will then amplify (we will see how this addition is done in the next sections). Our interest, then, is to model the transistor's response to this AC variation around the DC operating point. In what follows, it is important to keep in mind that we are dealing only with this AC response of the transistor; we have already solved the DC part of the

Figure 4.15 AC equivalent for the base-emitter junction.

problem when we determined the operating point. To emphasize this, we will use lower-case letters for AC currents and voltages.

First we consider the base-emitter junction. We assume the base current is some function of the base-emitter voltage, $I_b(V_{be})$. Since we are interested in variations around the DC operating point, we Taylor expand this function around the DC base-emitter voltage V_{be}^{DC}:

$$I_b(V_{be}) = I_b\left(V_{be}^{DC}\right) + \left(V_{be} - V_{be}^{DC}\right)\frac{dI_b}{dV_{be}}\left(V_{be}^{DC}\right) + \cdots. \tag{4.19}$$

The difference between the total current (or voltage) and the DC part of the current (or voltage) is the AC part of the current (or voltage). If this AC part is small,[1] we can truncate the series after the linear term, and Eq. (4.19) becomes

$$i_{be} = v_{be}\frac{dI_b}{dV_{be}}\left(V_{be}^{DC}\right) \tag{4.20}$$

where we have introduced the lower-case notation for the AC part of the signals. Finally, we define

$$r_{be} \equiv \left[\frac{dI_b}{dV_{be}}\left(V_{be}^{DC}\right)\right]^{-1} \tag{4.21}$$

so that Eq. (4.20) can be written

$$v_{be} = i_{be}r_{be}. \tag{4.22}$$

We have cast Eq. (4.22) in this form so that the interpretation will be clear: the AC or small signal response of the transistor can be modeled by a resistor with value given by Eq. (4.21) connected between the base and the emitter. This is represented in Fig. 4.15.

Note that, if we treat the base-emitter junction as a diode, we can evaluate Eq. (4.21) explicitly. In this case

$$I_b = I_0\left(e^{eV_{be}/kT} - 1\right) \tag{4.23}$$

---

[1] Because of this assumption, the model we are developing is sometimes called the *small signal model* or the *small signal equivalent* for the transistor

so

$$\frac{dI_b}{dV_{be}} = \frac{e}{kT}I_0 e^{eV_{be}/kT} \approx \frac{e}{kT}I_b \tag{4.24}$$

where, in this last relation, we have used the fact that, for any appreciable voltage, the exponential in Eq. (4.23) dominates. Since we are interested in evaluating the derivative at the DC operating point ($V_{be} \approx 0.6\,\mathrm{V}$), this approximation is valid. Thus

$$r_{be} = \frac{kT}{eI_b}. \tag{4.25}$$

Using room temperature for T gives

$$r_{be}(\Omega) = \frac{0.025}{I_b(A)}. \tag{4.26}$$

We next attempt to model the collector-emitter portion of the transistor. In this case, the collector current is a function of both the base current and the collector-emitter voltage, $I_c(I_b, V_{ce})$, so a double Taylor expansion is needed around the DC operating point:

$$I_c(I_b, V_{ce}) = I_c\left(I_b^{DC}, V_{ce}^{DC}\right) + \left(I_b - I_b^{DC}\right)\frac{dI_c}{dI_b}\left(I_b^{DC}, V_{ce}^{DC}\right)$$

$$+ \left(V_{ce} - V_{ce}^{DC}\right)\frac{dI_c}{dV_{ce}}\left(I_b^{DC}, V_{ce}^{DC}\right) + \cdots. \tag{4.27}$$

Introducing AC quantities as before and noting that $I_c = \beta I_b$ for the linear active region, we obtain

$$i_c = \beta i_b + \frac{1}{r_{out}}v_{ce} \tag{4.28}$$

where

$$r_{out} \equiv \left[\frac{dI_c}{dV_{ce}}\left(I_b^{DC}, V_{ce}^{DC}\right)\right]^{-1} \tag{4.29}$$

and the β in Eq. (4.28) is obtained at the operating point. The circuit equivalent for Eq. (4.28) is shown in Fig. 4.16. A current source with value βi_b is in parallel with a resistor r_{out}. A voltage v_{ce} applied across the terminals will give a current v_{ce}/r_{out} through the resistor, which, when combined with the current source, gives the collector current expressed by Eq. (4.28).

Putting both parts of the AC equivalent model together yields Fig. 4.17. The left side of the figure shows the results of Figs. 4.15 and 4.16 connected together at the emitter. This can be redrawn in the compact form shown on the right.

Bipolar junction transistors

Table 4.1 Comparison of AC transistor models

Transistor parameter	Our model	h-parameter
Input resistance	r_{be}	h_{ie}
Output resistance	r_{out}	$\frac{1}{h_{oe}}$
Current gain	β	h_{fe}
Voltage feedback ratio	none	h_{re}

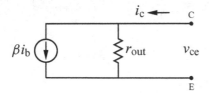

Figure 4.16 AC equivalent for the collector-emitter.

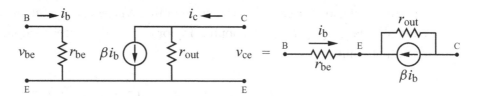

Figure 4.17 The completed AC equivalent for the transistor.

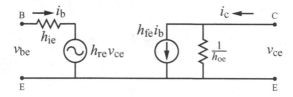

Figure 4.18 The h-parameter model for the transistor.

Before moving on, we note that more complicated models of transistor behavior exist. One of these, the *h-parameter model*, is shown in Fig. 4.18. In addition to some name changes (summarized in Table 4.1), this model adds a voltage source $h_{re}v_{ce}$ to the base-emitter junction. This addition recognizes that the base-emitter junction is not an independent diode, but is part of a transistor, and the properties of the base-emitter junction will depend on the voltage applied to the collector-emitter junction. Note that there is potential for confusion in the transistor parameter names. The terms *input resistance, output resistance,* and *current gain* here refer to the *transistor*. We also apply these same terms to *amplifiers*. Fortunately, the symbols used in this latter case are different (r_{be}, r_{out}, and β versus Z_{in}, Z_{out}, and g).

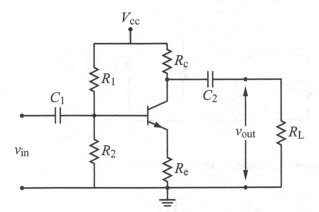

Figure 4.19 The common-emitter amplifier.

4.4.4 Applying the AC equivalents: the common-emitter amplifier

We now apply the AC equivalents to the calculation[2] of our four amplifier parameters a, g, Z_{in}, and Z_{out}. We will do this for three different amplifier circuit configurations. The first of these is the common-emitter amplifier, shown in Fig. 4.19.

Since this is the first of our amplifier configurations, some comments are in order. Note that the central part of this circuit is the universal DC bias circuit which, as we have seen, sets the DC operating point for the circuit. We now add an AC signal v_{in} to the mix, coupled to the transistor base through the capacitor C_1. For this configuration, we take our output voltage off the collector of the transistor, coupling this through capacitor C_2. A resistor R_L representing the load is included on the output. The coupling capacitors C_1 and C_2 insure that the DC operating point will not be affected by the circuitry connected to v_{in} and v_{out}.

The AC analysis of the amplifier can be broken down into steps as follows. It is usually helpful to redraw the circuit as one proceeds.

1. Treat the coupling capacitors C_1 and C_2 as shorts and the DC power supply as ground. Here we make the approximation that the coupling capacitance is large enough that the capacitive impedance is negligible. Similarly, power supplies usually have a large capacitance to ground, so the AC signal is effectively connected to ground.

2. Insert the AC transistor model and simplify the circuit as much as possible. It is important to keep track of various quantities of interest by including them in

---

[2] We will calculate the voltage gain a including a load resistor R_L to simplify the calculation of the current gain. The black box parameter a_{OL} can be obtained from a by setting $R_L = \infty$.

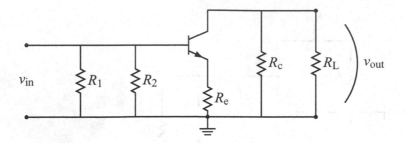

Figure 4.20 The common-emitter amplifier after applying step 1.

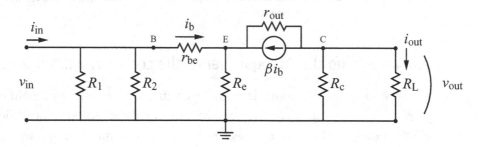

Figure 4.21 The common-emitter amplifier after inserting the transistor model.

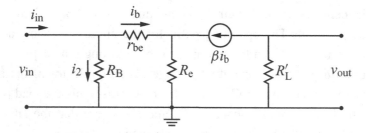

Figure 4.22 The common-emitter amplifier after simplifying.

the redrawn circuit. These include v_{in}, v_{out}, i_{in}, i_{out}, i_{b}, and the location of the transistor base, emitter, and collector.

3. Compute a, g, Z_{in}, and Z_{out} by writing v_{in}, v_{out}, i_{in}, and i_{out} in terms of i_{b}.

Applying step 1 to the common-emitter amplifier, we obtain the redrawn circuit shown in Fig. 4.20. The capacitors are replaced by direct connections and resistors R_1 and R_{c}, which were originally connected to V_{cc}, are connected to ground. Notice that we have been careful to label v_{in} and v_{out} on the redrawn circuit.

Applying step 2, we obtain the redrawn circuit shown in Figs. 4.21 and 4.22. We have inserted the transistor equivalent of Fig. 4.17, being careful to label everything as we go. To obtain the simplified version shown in Fig. 4.22, we replace the parallel combination of R_1 and R_2 by R_B and the parallel combination of R_c and R_L by R'_L. Finally, we note that typically r_{out} is large and, to a first approximation, can be ignored, so we remove it from the circuit.

We are now ready to apply step 3. The trick here is to write everything in terms of i_b, which will then cancel out in the end. For example, the output voltage, v_{out}, is equal to $-\beta i_b R'_L$ since the current from the current source must go through R'_L. The minus sign reflects the fact that, due to the direction of the current, we have a voltage drop. To obtain the input voltage, v_{in}, we add the voltage across r_{be} to the voltage across R_e, giving $v_{in} = i_b r_{be} + (i_b + \beta i_b) R_e$. Hence the voltage gain is

$$a = \frac{v_{out}}{v_{in}} = \frac{-\beta i_b R'_L}{i_b[r_{be} + (\beta + 1)R_e]} = \frac{-\beta R'_L}{r_{be} + (\beta + 1)R_e}. \tag{4.30}$$

Note that in this last result i_b has canceled out and we are left with a result that depends only on circuit parameters that we know: r_{be}, β, and the resistor values.

Turning next to the current gain, we need expressions for i_{out} and i_{in}. Note here that i_{out} is the current through the load resistor R_L, not R'_L (cf. Fig. 4.21). Since we already have an expression for v_{out}, it is easiest to note that $i_{out} = v_{out}/R_L$. For i_{in}, note from Fig. 4.22 that $i_{in} = i_2 + i_b$ and $i_2 = v_{in}/R_B$. Again, we can use the previously obtained expression for v_{in} to complete this:

$$g = \frac{i_{out}}{i_{in}} = \frac{\frac{v_{out}}{R_L}}{i_b\left[\frac{r_{be}+(\beta+1)R_e}{R_B} + 1\right]} = \frac{-\beta i_b \frac{R_c}{R_c+R_L}}{i_b \frac{R_B+r_{be}+(\beta+1)R_e}{R_B}}$$

$$= -\beta\left(\frac{R_B}{R_B + r_{be} + (\beta + 1)R_e}\right)\left(\frac{R_c}{R_c + R_L}\right). \tag{4.31}$$

Finally, we compute the input and output impedances, using $Z_{in} = v_{in}/i_{in}$ and Eq. (4.16). Again, we can use the expressions for v_{in}, v_{out}, i_{in}, and i_{out} we have already obtained:

$$Z_{in} = \frac{v_{in}}{i_{in}} = \frac{i_b[r_{be} + (\beta + 1)R_e]}{i_b\left[\frac{R_B+r_{be}+(\beta+1)R_e}{R_B}\right]} = \frac{R_B[r_{be} + (\beta + 1)R_e]}{R_B + r_{be} + (\beta + 1)R_e} \tag{4.32}$$

and

$$Z_{out} = \frac{v_{out}(R_L = \infty)}{i_{out}(R_L = 0)} = \frac{-\beta i_b R_c}{-\beta i_b} = R_c. \tag{4.33}$$

We will examine the meaning of these results when we compare our three amplifiers at the end. For now we simply note again that our results depend only on the known circuit parameters, so we can, in principle, adjust the values to match our requirements.

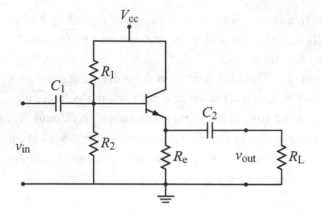

Figure 4.23 The common-collector amplifier.

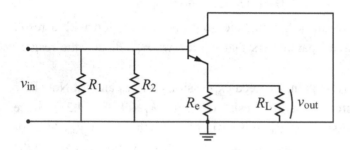

Figure 4.24 The common-collector amplifier after applying step 1.

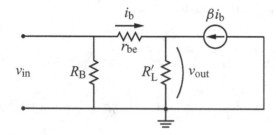

Figure 4.25 The common-collector amplifier after inserting the transistor model and simplifying.

4.4.5 The common-collector amplifier

Our next amplifier configuration, called the *common-collector amplifier*,[3] is shown in Fig. 4.23. In this case the output is taken from the emitter of the transistor, and the collector resistor is omitted. Our DC operating point formulas are still valid, but we must set $R_c = 0$.

Following the steps in our analysis recipe, we obtain the versions of this circuit shown in Figs. 4.24 and 4.25. As before, we have combined R_1 and R_2 into R_B, R_e and R_L into R_L', and ignored r_{out}.

---

[3] This configuration is also called the *emitter-follower*.

The voltage gain a is then given by

$$a = \frac{v_{\text{out}}}{v_{\text{in}}} = \frac{i_b(\beta + 1)R'_L}{i_b r_{be} + i_b(\beta + 1)R'_L} = \frac{(\beta + 1)R'_L}{r_{be} + (\beta + 1)R'_L}. \qquad (4.34)$$

The current gain g and the impedances Z_{in} and Z_{out} are left as exercises, but the results are

$$g = (\beta + 1) \left(\frac{R_B}{R_B + r_{be} + (\beta + 1)R'_L} \right) \left(\frac{R_e}{R_e + R_L} \right), \qquad (4.35)$$

$$Z_{\text{in}} = \frac{R_B[r_{be} + (\beta + 1)R'_L]}{R_B + r_{be} + (\beta + 1)R'_L}, \qquad (4.36)$$

and

$$Z_{\text{out}} = R_e. \qquad (4.37)$$

4.4.6 The common-base amplifier

Our final amplifier configuration, the *common-base amplifier*, is shown in Fig. 4.26. Here the input voltage is applied to the emitter, the output is taken from the collector, and an additional capacitor connects the base to ground. Note again that the central portion of this circuit is the same DC bias circuit common to all three amplifiers.

Our analysis steps give the versions of the circuit shown in Figs. 4.27 and 4.28. Because the capacitor C_B shorts the transistor base to ground, resistors R_1 and R_2 drop out of the final drawing. We also combine R_c and R_L to form R'_L, and in Fig. 4.28 we redraw the circuit one last time to make the calculations more transparent.

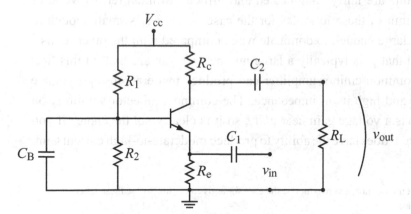

Figure 4.26 The common-base amplifier.

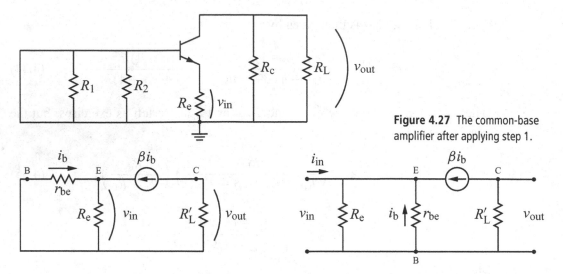

Figure 4.27 The common-base amplifier after applying step 1.

Figure 4.28 The common-base amplifier after inserting the transistor model and simplifying. The two versions shown are equivalent but the one on the right makes the calculations easier.

The details of the calculations are left as exercises. The results are

$$a = \beta \frac{R'_L}{r_{be}}, \tag{4.38}$$

$$g = \beta \left(\frac{R_e}{r_{be} + (\beta + 1)R_e} \right) \left(\frac{R_c}{R_c + R_L} \right), \tag{4.39}$$

$$Z_{in} = \frac{r_{be} R_e}{r_{be} + (\beta + 1)R_e}, \tag{4.40}$$

and

$$Z_{out} = R_c. \tag{4.41}$$

The results for our three transistor amplifier circuits are summarized in Table 4.2. Some of our results are fairly complicated and difficult to interpret, so we have also shown the limit of these formulas for the case where r_{be} is small enough to ignore and R_B is large enough to dominate when compared with the other terms.[4] Keeping in mind that β is typically a large number, we can see that, in this limiting case, the common-emitter amplifier can produce moderate-to-high voltage and current gain and high input impedance. The common-collector amplifier, on the other hand, has a voltage gain near unity, so it is clearly not the choice if you want voltage gain. It does have the ability to produce moderate-to-high current gain,

---

[4] This situation is not uncommon, but may not be the case for a given circuit, so use these latter expressions with care.

Table 4.2 Summary of results for the bipolar transistor amplifiers

	Common-emitter	Common-collector	Common-base
a	$\dfrac{-\beta(R_c\|R_L)}{r_{be}+(\beta+1)R_e}$ $\to -\left(\dfrac{\beta}{\beta+1}\right)\dfrac{R_c\|R_L}{R_e}$	$\dfrac{(\beta+1)R_0}{r_{be}+(\beta+1)R_0}\to 1$	$\dfrac{\beta(R_c\|R_L)}{r_{be}}$
g	$\dfrac{-\beta R_B}{R_B+r_{be}+(\beta+1)R_e}\left(\dfrac{R_c}{R_c+R_L}\right)$ $\to -\beta\left(\dfrac{R_c}{R_c+R_L}\right)$	$\dfrac{(\beta+1)R_B}{R_B+r_{be}+(\beta+1)R_0}\left(\dfrac{R_0}{R_L}\right)$ $\to (\beta+1)\dfrac{R_e}{R_e+R_L}$	$\dfrac{\beta R_e}{r_{be}+(\beta+1)R_e}\left(\dfrac{R_c}{R_c+R_L}\right)$ $\to \dfrac{\beta}{\beta+1}\left(\dfrac{R_c}{R_c+R_L}\right)$
Z_{in}	$\dfrac{R_B[r_{be}+(\beta+1)R_e]}{R_B+r_{be}+(\beta+1)R_e}$ $\to (\beta+1)R_e$	$\dfrac{R_B[r_{be}+(\beta+1)R_0]}{R_B+r_{be}+(\beta+1)R_0}$ $\to (\beta+1)R_0$	$\dfrac{r_{be}R_e}{r_{be}+(\beta+1)R_e}\to \dfrac{r_{be}}{\beta+1}$
Z_{out}	R_c	R_e	R_c

$R_c\|R_L$ is the parallel combination of R_c and R_L.
$R_B = R_1\|R_2$.
$R_0 = R_E\|R_L$.
The arrow shows the limit when r_{be} is small and R_B is large.

moderate-to-high input impedance, and low output impedance. These latter two features are the strengths of this configuration for they allow this circuit to act as a buffer between a high impedance source and a low impedance load. If such a source was connected directly to the load, the voltage across the load would be severely reduced by the voltage-divider effect and the power transfered to the load would be small. One can alleviate these effects by using this amplifier between the source and the load.

The common-base amplifier can produce very high voltage gain since β is large and r_{be} can be small. The current gain, in contrast, will always be less than one and the input impedance will be very low. Since these latter properties are usually not desirable, this amplifier is only used when one is willing to pay the price for the high voltage gain.

4.4.7 Other properties of transistor amplifiers

While we have thus far treated amplifier gain as a constant, it is really a function of input signal frequency. One source of this frequency dependence is the capacitors in the amplifier circuits. Recall that we have assumed in our derivations that these are large enough that we can ignore the capacitive impedance $Z_c = 1/(j\omega C)$. If we make ω small enough, however, this impedance will become large and the amplifier gain will decrease as a result.

There is also a decrease in amplifier gain at high frequency. This effect is not due to the coupling capacitors since, for high frequencies, our zero-impedance approximation is very good. Instead, this decrease is due to so-called *stray capacitances* in the circuit. Generally speaking, there is capacitance between any two conductors. This means that the wires that connect the components and the leads of the components have a small capacitance between them. Usually this capacitance is small enough to ignore (just as we ignore the resistance of these wires), but as the signal frequency increases the capacitive impedance due to these stray capacitances becomes small enough to make a difference. There is also stray capacitance at the semiconductor junctions; the charge separation that results in energy level shifts makes the junction look like a capacitor with positive charge on one side of the junction and negative charge on the other. The result of all this stray capacitance is that high frequency amplifier signals have a low impedance to ground, and this reduces the amplifier gain at high frequencies.

These effects are usually summarized by giving the *frequency response* of the amplifier. Typical parameters involved in this specification are included in Fig. 4.29. The voltage gain is plotted versus frequency on a log-log plot. The constant gain

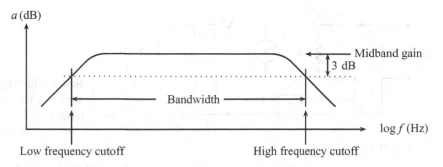

Figure 4.29 An example frequency response curve for an amplifier.

level for the middle frequencies is termed the *midband gain*. The decrease in the gain at low and high frequencies is called the *roll off* of the gain. Since the gain decreases in a continuous manner, we choose a level 3 dB below the midband gain to define the *low frequency* and *high frequency cutoffs* and the *bandwidth*. In amplifier specifications, these quantities are often given in a table rather than as a graph as we have done.

Finally, we note that these same capacitive effects give rise to frequency-dependent *phase shifts* between the input and output signals. This behavior is familiar from our early work on RC circuits where the output had a phase that was dependent on ω. The same thing happens in any circuit with a capacitance (or inductance) since the impedance of these components is imaginary.

4.4.8 Distortion

Ideally, our amplified signal will simply be a larger version of our input signal, but in practice there is always some *distortion* of the signal. We saw a rather severe example of this in Fig. 4.8, where the sine-wave output was clipped off when the input signal became negative. We addressed this problem by placing our DC operating point in the center of the linear active region where the relationship between the base and collector currents is roughly linear. It is not perfectly linear, however, and the non-linearity becomes progressively worse as the amplitude of the signal into the base increases. Eventually, the output will show signs of clipping at one or both extremes as the transistor is driven into the saturation or cutoff regimes. This type of distortion is called *harmonic distortion* because a Fourier analysis of the distorted output signal will include frequencies that are integer multiples of the pure sine-wave input.

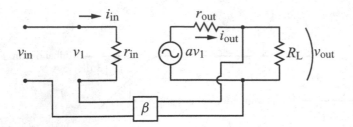

Figure 4.30 A black box amplifier with voltage feedback.

4.4.9 Feedback

We can obtain an interesting and useful modification of our amplifier properties by adding *feedback* to the circuit. This is done by taking a fraction β of the output voltage v_{out} and adding it to the input voltage.[5] This is shown schematically in Fig. 4.30 where we have modified the usual black box model of an amplifier to include the feedback. We imagine that the circuitry in the box marked β does the feedback job for us, and we assume this box has a large enough impedance on the right side that it does not affect the output voltage or current. We can then still write

$$v_{out} = av_1 - i_{out}r_{out} \qquad (4.42)$$

for the output loop. On the input side, we have

$$v_{in} = v_1 - \beta v_{out}. \qquad (4.43)$$

Eliminating v_1 from these equations gives

$$v_{out} = av_{in} + a\beta v_{out} - i_{out}r_{out} \qquad (4.44)$$

and solving for v_{out} yields

$$v_{out} = \left(\frac{a}{1 - a\beta}\right)v_{in} - \left(\frac{r_{out}}{1 - a\beta}\right)i_{out}. \qquad (4.45)$$

We now notice that Eq. (4.45) has the same form as Eq. (4.42). This means that the new circuit, which includes the feedback loop, can be viewed as a new amplifier with a modified voltage gain a' and output resistance r'_{out} given by

$$a' = \left(\frac{a}{1 - a\beta}\right) \qquad (4.46)$$

and

$$r'_{out} = \left(\frac{r_{out}}{1 - a\beta}\right). \qquad (4.47)$$

[5] Note that this β is not the same as the transistor current gain.

Similar arguments can be made to show that the input resistance of our new amplifier is also modified. Applying KVL to the output gives

$$v_{out} = i_{out}R_L = av_1 - i_{out}r_{out} \qquad (4.48)$$

which can be solved for i_{out}:

$$i_{out} = \frac{av_1}{R_L + r_{out}}. \qquad (4.49)$$

The input loop gives

$$v_1 = v_{in} + \beta v_{out} = v_{in} + \beta i_{out}R_L = v_{in} + \beta R_L \left(\frac{av_1}{R_L + r_{out}}\right) v_1. \qquad (4.50)$$

Rearranging, we obtain

$$v_1 \left(1 - \frac{a\beta}{1 + \frac{r_{out}}{R_L}}\right) = v_{in}. \qquad (4.51)$$

Finally, we employ $v_1 = i_{in}r_{in}$ in Eq. (4.51) and note that the new input resistance r'_{in} is just v_{in}/i_{in}. Hence

$$\frac{v_{in}}{i_{in}} \equiv r'_{in} = \left(1 - \frac{a\beta}{1 + \frac{r_{out}}{R_L}}\right) r_{in}. \qquad (4.52)$$

Equations (4.46), (4.47), and (4.52) show how the amplifier voltage gain, output resistance, and input resistance are modified by the addition of feedback. The derivation has been general, so we can now consider special cases. Note first that if $a\beta = 1$, the modified gain $a' \to \infty$, so even the smallest input voltage will be amplified until the circuitry reaches its limits.[6] This is the basis of oscillator circuits, which we will examine in detail later.

A particularly interesting limit of these general expressions is the case where $a\beta$ is large and negative. In this case

$$a' = \left(\frac{a}{1 - a\beta}\right) \approx \left(\frac{a}{-a\beta}\right) = -\frac{1}{\beta} \qquad (4.53)$$

so the modified voltage gain a' in this limit depends only on the feedback β and not on the original amplifier gain a. Similarly,

$$r'_{out} = \left(\frac{r_{out}}{1 - a\beta}\right) \approx \left(\frac{r_{out}}{-a\beta}\right) \to 0 \qquad (4.54)$$

---

[6] This case is well known to most garage bands.

and

$$r'_{in} = \left(1 - \frac{a\beta}{1 + \frac{r_{out}}{R_L}}\right) r_{in} \approx \left(-\frac{a\beta}{1 + \frac{r_{out}}{R_L}}\right) r_{in} \to \infty \qquad (4.55)$$

where the final limits pertain to the case $-a\beta \to \infty$.

These characteristics are quite attractive. The input resistance is high, so we need not worry about having our source voltage loaded down by the voltage divider effect. A similar advantage is obtained on the output because the output resistance is low. Finally, the voltage gain is set by a single parameter, the feedback β, in contrast to the relatively complicated expressions for voltage gain in Table 4.2. Appreciation of the characteristics of amplifiers with this type of feedback led to the development of the *operational amplifier* which we will study in Chapter 6.

EXERCISES

1. Using the transistor characteristics of Fig. 4.31, find β for several values of I_b when $V_{ce} = 6$ V. Repeat for several values of V_{ce} when $I_b = 30$ μA. This shows that β is not really a constant over the linear active region.

2. Determine the operating point of a universal transistor DC bias circuit when $V_{cc} = 15$ V, $R_1 = 10$ kΩ, $R_2 = 2.2$ kΩ, $R_c = 680$ Ω, and $R_e = 100$ Ω. Assume $\beta = 200$ and $V_{be} = 0.72$ V.

3. Design a circuit that will set a reasonable operating point for a transistor with the characteristics of Fig. 4.31. Assume that the power rating for the transistor is 25 mW.

4. For the operating point of the previous problem, determine the AC model parameters r_{be}, β, and r_{out}.

5. Derive the expressions for the current gain g and the input impedance Z_{in} for the common-collector amplifier.

6. Derive the expressions for the voltage gain a, the current gain g and the input impedance Z_{in}, and the output impedance Z_{out} for the common-base amplifier.

7. If the circuit of Problem 2 is configured as a common-emitter amplifier, calculate the resulting voltage and current gain. Assume a load resistor of 1 kΩ.

8. Design a transistor amp with $a \approx 1$ and $g \approx 100$. Give the values of all components you use. Assume the transistor has the characteristics of Fig. 4.31. You may assume any load resistance you like.

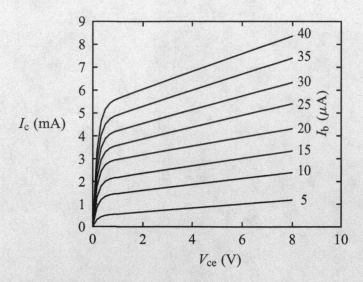

Figure 4.31 Transistor I–V characteristics for Problems 1, 3, 4, and 8.

9. Suppose an amplifier has an open-loop voltage gain of 20, an input impedance of 100 Ω, and an output impedance of 50 Ω. The amplifier is driven with a sine-wave generator with output impedance of 50 Ω and an open loop amplitude of 0.1 V_{pp}. Find the resulting voltage across a 200 Ω load attached to the amplifier output.

FURTHER READING

David Casasent, *Electronic Circuits* (New York: Quantum, 1973).

Richard C. Jaeger and Travis N. Blalock, *Microelectronic Circuit Design*, 3rd edition (New York: McGraw-Hill, 2008).

Donald A. Neamen, *Microelectronics: Circuit Analysis and Design*, 3rd edition (New York: McGraw-Hill, 2007).

John E. Uffenbeck, *Introduction to Electronics, Devices and Circuits* (Englewood Cliffs, NJ: Prentice-Hall, 1982).

5 Field-effect transistors

5.1 Introduction

In this chapter we introduce the second major type of transistor: the *field-effect transistor*. Like the bipolar junction transistors (BJTs) we studied in Chapter 4, field-effect transistors (FETs) allow the user to control a current with another signal. The key difference is that the FET control signal is a voltage while the BJT control signal is a current. Also, the FET control input (called the gate) has a much higher input impedance than the base of a BJT. Indeed, the DC gate impedance for FETs varies from a few megaohms to astounding values in excess of 10^{14} Ω. High input impedance is a highly desirable feature that greatly simplifies circuit analysis.

The BJT has three connections: the collector, base, and emitter. The corresponding connections on an FET are called the *drain*, *gate*, and *source*. Some versions of the FET have a fourth connection called the *bulk* connection. Bipolar transistors come in just two types with opposite polarities: the npn and the pnp. Field-effect transistors have greater variety. In addition to the polarity pairs (termed *n-channel* and *p-channel*), there are differences in gate construction (*junction* and *metal oxide*), and doping (*depletion* and *enhancement*). In terms of analysis, however, they are all very similar, so we will not have to consider each variety separately. Also, as we did with the bipolar transistor, we will focus on one of the polarities (n-channel) since the other polarity simply involves swapping the labels for n and p and changing the sign of the voltages and the direction of the currents.

FETs have both advantages and disadvantages when compared with BJTs. As noted above, FETs have extremely high gate impedance which makes them ideal as a buffer or input stage for a complex circuit. They are also generally less sensitive to temperature variations and more suitable for the large scale integration of modern micro-circuits. On the other hand, FET amplifiers tend to have lower gain than their BJT counterparts. The metal oxide gate construction of some FETs is highly susceptible to damage from static electricity, which means that you can destroy the transistor simply by touching it. Finally, the manufacturing spread in FET parameters is larger than for BJTs. This makes it trickier to design circuits for mass production.

5.2 Field-effect transistor fundamentals

5.2.1 Junction field-effect transistors

We first consider the *junction field-effect transistor* or JFET. The n-channel version of the JFET is shown with typical external bias voltages in Fig. 5.1. The gate of the device is a piece of p-type semiconductor placed in a larger piece of n-type semiconductor. The two ends of the n-type semiconductor are called the source and the drain, and the region surrounding the p-n junction is called the *channel*. Current enters the drain, flows through the channel and exits from the source. Since the current is flowing through a single material, the I–V characteristic between the drain current I_d and the drain-source voltage V_{ds} is that of a resistor (i.e., linear). However, this simple behavior is modified when a gate voltage is applied or when V_{ds} gets too large.

As shown in the figure, the polarity of the gate-source voltage V_{gs} is such as to reverse bias the p-n junction of this device. As noted in Chapter 3, a reverse biased p-n junction passes very little current, and this is the reason for the high gate input impedance of this device. Recall also that there is a depletion region in the vicinity of the p-n junction where the density of the charge carriers is markedly reduced, and the size of this depletion region increases with reverse bias. Thus, as V_{gs} becomes more negative the depletion region extends further into the channel, effectively reducing its cross-sectional area. Since the resistance of a material varies inversely

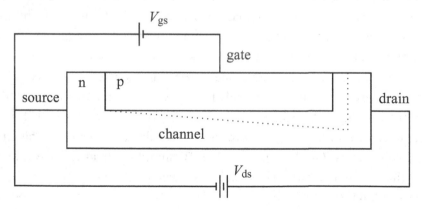

Figure 5.1 An n-channel JFET with typical biasing voltages.

with cross-sectional area (cf. Eq. (1.6) in Chapter 1), this reduces the drain current and the slope of the I–V characteristic curve.

This, however, is not the whole story. The potential of the channel relative to the source varies from zero to V_{ds} as we move thorough the channel from the source to the drain. Thus, the reverse bias of the p-n junction and the extent of the depletion region also vary with position. This is indicated in Fig. 5.1 by the dotted line representing the boundary of the depletion region.

If the reverse bias of the p-n junction at some location is sufficient, the depletion region will extend across the entire channel and the boundary will touch the bottom of the channel. This is called the *pinchoff* or *saturation* point. If V_{ds} is further increased after saturation, the drain current remains essentially constant, reflecting a balance between the increased voltage and the reduced conductivity of the channel.

A set of representative I–V characteristics showing these features is given in Fig. 5.2, which also serves to introduce some additional nomenclature. The different curves represent different values of V_{gs} which are all taken to be above the *threshold value* V_t, the value of V_{gs} that reduces the drain current to zero for all V_{ds}. For small V_{ds} the curves are linear, but eventually curve over and reach their maximum value $I_{d(sat)}$ at $V_{ds(sat)}$. The boundary of this *linear region* (also called the *resistance, ohmic,* or *non-saturation* region) is indicated by the dotted line. For values of $V_{ds} > V_{ds(sat)}$, the drain current is essentially constant. This is called the *saturation*

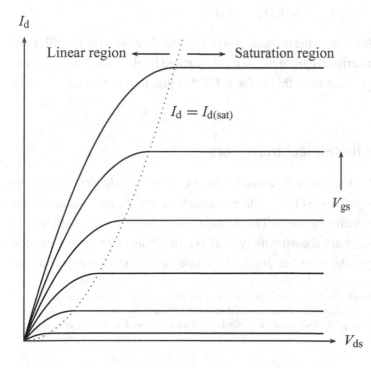

Figure 5.2 I–V characteristics for an FET.

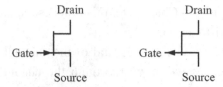

Figure 5.3 Circuit symbols for the n-channel (left) and p-channel (right) JFETs.

region (also known as the *pinchoff* or *active* region). Although not shown in Fig. 5.2, for large enough V_{ds} we enter a *breakdown region* where the drain current rises steeply.

The characteristic curves can also be represented by a set of model equations.[1] The saturation value of the drain-source voltage for a given gate-source voltage is given by

$$V_{ds(sat)} = V_{gs} - V_t. \tag{5.1}$$

Here V_t is the threshold voltage previously defined. In the linear region where $V_{ds} < V_{ds(sat)}$, the drain current is given by

$$I_d = K[2(V_{gs} - V_t) - V_{ds}]V_{ds} = K[2V_{ds(sat)} - V_{ds}]V_{ds} \tag{5.2}$$

where K is a constant. From this we can see that an approximately linear relationship between I_d and V_{ds} requires $V_{ds} \ll 2V_{ds(sat)}$. Finally, in the saturation region where $V_{ds} > V_{ds(sat)}$, we have

$$I_d = I_{d(sat)} = K(V_{gs} - V_t)^2 = KV_{ds(sat)}^2. \tag{5.3}$$

The circuit symbols for both the n-channel and p-channel versions of the JFET are shown in Fig. 5.3. The arrow represents the diode formed by the transistor junction. One must keep in mind, however, that in the JFET this junction is normally reverse biased.

5.2.2 Metal oxide field-effect transistors

The second type of FET we will consider is the *metal oxide semiconductor field-effect transistor* or MOSFET.[2] These transistors are further divided into *enhancement* and *depletion* versions. The n-channel versions of the enhancement and depletion MOSFETs are shown with typical external bias voltages in Figs. 5.4 and 5.5, respectively.[3] As with the JFET, the name of this device refers to the

---

[1] We refer the reader to the references for further discussion of the physical basis of these equations.
[2] This device is sometimes called an IGFET for *insulated gate field-effect transistor*.
[3] In some advanced applications, the bulk connection would have its own bias, but we will assume it is connected to the source

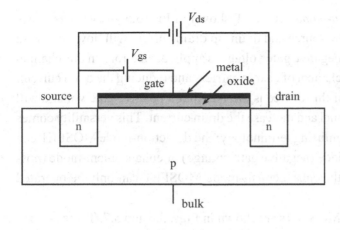

Figure 5.4 An n-channel enhancement-mode MOSFET with typical biasing voltages.

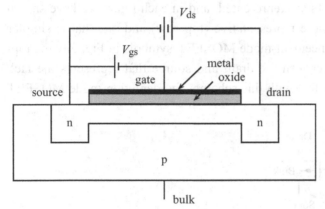

Figure 5.5 An n-channel depletion-mode MOSFET with typical biasing voltages.

structure of the gate. For the MOSFET, the gate pin of the device connects to a layer of metal or other high conductivity material. The metal is deposited on a layer of silicon oxide or other insulating material, which in turn is deposited on the semiconductor. Because of the insulating material, the gate is electrically isolated from the semiconductor and has an extremely high DC input impedance.

To understand the operation of MOSFETs, consider Fig. 5.4. The source and drain are connected to two pieces of n-semiconductor embedded in a p-semiconductor substrate (bulk). Without a gate voltage, the source-bulk and drain-bulk junctions form two back-to-back diodes and no current can flow from drain to source. When a positive voltage is applied to the gate, minority electrons in the p-semiconductor are attracted to the gate region and form an *inversion layer*. The electrons in this inversion layer allow current to flow between the drain and source. The larger the gate voltage, the more electrons in the inversion layer and the larger the drain current. This type of MOSFET, where a conduction channel is created by the action of the gate voltage, is called an *enhancement-mode* MOSFET.

In contrast, the *depletion-mode* MOSFET shown in Fig. 5.5 already has a permanent channel between the source and drain, so drain current will flow even with zero gate voltage. When a negative gate voltage is applied, electrons in this channel are repelled, depleting the channel of charge carriers and reducing the drain current. Adding to the versatility of this device is the fact that a positive gate voltage will attract electrons to the channel and increase the drain current. This versatility comes at the price of somewhat confusing terminology: the depletion-mode MOSFET can be operated in depletion-mode (negative gate voltage) or enhancement-mode (positive gate voltage), while the enhancement-mode MOSFET can only be operated in enhancement-mode.

The circuit symbols for MOSFETs are shown in Figs. 5.6 and 5.7. There is some variety in the way MOSFETs are represented, and in each figure we have shown two versions, one that is more representative (top row) and one that is simpler (bottom row). For the enhancement-mode MOSFET symbols in Fig. 5.6, the top-row versions have a gap between the drain and source that represents the lack of a permanent channel. In line with this scheme, the depletion-mode MOSFET

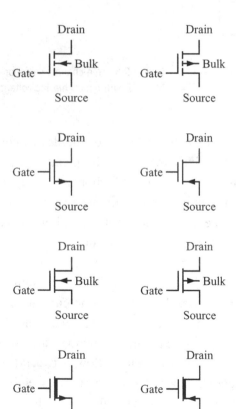

Figure 5.6 Circuit symbols for the n-channel (left) and p-channel (right) enhancement-mode MOSFETs. The second row shows simplified versions of the symbols.

Figure 5.7 Circuit symbols for the n-channel (left) and p-channel (right) depletion-mode MOSFETs. The second row shows simplified versions of the symbols.

symbols in Fig. 5.7 have a solid line between drain and source. Some MOSFETS have a separate pin for the bulk connection (also called the body or substrate connection) and this is shown in the top-row symbols. The bottom-row symbols have the advantage of being similar to those for bipolar transistors and reflect the fact that the bulk connection is often simply tied to the source connection.

The variety of field-effect transistors can be daunting, but the various types are actually quite similar in terms of circuit analysis. The I–V characteristics in Fig. 5.2 could apply to any of the FETs we have discussed, the only difference being the values of V_{gs} assigned to the curves. For the n-channel JFET, the threshold voltage is always negative and there is not much to gain by making the gate voltage greater than zero. For the n-channel enhancement-mode MOSFET, the threshold voltage is zero and the higher curves would correspond to increasingly positive values of V_{gs}. For the n-channel depletion-mode MOSFET, the threshold voltage is negative and a zero value of V_{gs} would correspond to one of the mid-level curves. Higher curves would be positive values of V_{gs}.

A graphical way to see this is to plot the *transfer curve* for the various devices. The transfer curve is a plot of I_d versus V_{gs} for a fixed value of V_{ds} in the saturation region. In Fig. 5.8 we have plotted representative transfer curves for the three types

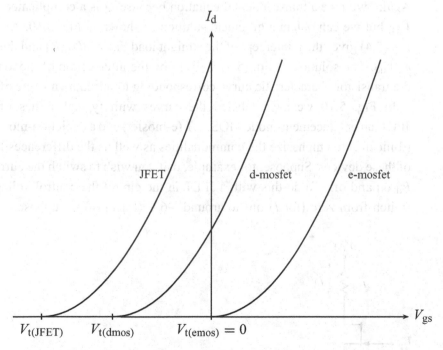

Figure 5.8 Transfer curves for (from left to right) a JFET, a depletion-mode MOSFET, and an enhancement-mode MOSFET (all n-channel).

of FET we have discussed. The functional dependence is given by Eq. (5.3). The reader can verify that the curves reflect the qualitative description given above.

5.3 DC and switching applications

As we did with the BJT, we first consider the use of an FET to provide either DC or on-off control of a current. The relevant circuit is shown using a JFET in Fig. 5.9, but the analysis is the same for a MOSFET. Here V_{dd} is a constant power supply voltage and V_g is the control voltage, which controls the flow of current through resistor R_d. As before, both voltages are understood to be relative to ground.

The analysis of the control circuit is simplified by the fact that FETs are voltage controlled devices. We only need find the voltage between the gate and source V_{gs}, and, since the source is grounded, $V_{gs} = V_g$. The drain circuit analysis gives $V_{dd} - I_d R_d - V_{ds} = 0$, where V_{ds} is the voltage from the drain to the source of the FET. Solving for I_d gives

$$I_d = \frac{V_{dd} - V_{ds}}{R_d}. \tag{5.4}$$

Again, we face a transcendental equation because I_d is a complicated function of V_{ds}, but we can obtain a graphical solution as shown in Fig. 5.10. An analysis of Eq. (5.4) gives the y-intercept of the straight load line as V_{dd}/R_d and the x-intercept as V_{dd}. The solution of Eq. (5.4) is given by the intersection of the load line with the transistor characteristic curve corresponding to our known value of V_{gs}.

In Fig. 5.10 we have labeled the curves with typical values of V_{gs} for a JFET, an enhancement-mode MOSFET (e-mosfet) and a depletion-mode MOSFET (d-mosfet) to emphasize the commonalities as well as the differences in the usage of these devices. Suppose, for example, that you wish to switch the current through R_d on and off. To do this with a JFET in the circuit the control voltage V_g must switch from zero (for I_d on) to around $-6\,$V (for I_d off). Suppose now that you

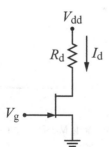

Figure 5.9 JFET switching circuit.

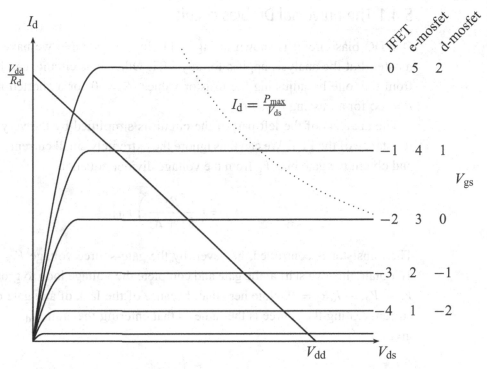

Figure 5.10 Graphical solution for the circuit of Fig. 5.9.

only have positive voltages available for V_g. The circuit can be adapted to this situation by simply using an e-mosfet rather than the JFET. Now +5 V would turn I_d on and 0 V would turn it off. Finally, we note that, as always, the circuit design is constrained by the power rating of the FET used and our operating point must be to the left of the maximum power curve shown by the dotted line.

5.4 Amplifiers

We now turn to the use of FETs in amplifier circuits. Again, the development is very similar to that used for BJT amplifier circuits in Chapter 4. The first task is to set the DC operating point of the transistor in the central part of the saturation region. We will use the universal DC bias circuit to accomplish this. Then we will develop an AC equivalent circuit to describe variations around the operating point. This equivalent circuit will then be used to analyze typical amplifier configurations and obtain the four quantities (a, g, Z_{in}, and Z_{out}) required for the black box model of the amplifiers.

5.4.1 The universal DC bias circuit

Our DC bias circuit is shown in Fig. 5.11. In this example we have used an e-mosfet, but the analysis applies for any FET. Other bias circuits can be obtained from this one by adjusting the resistor values ($R = 0$ for a shorted resistor and $R = \infty$ for a missing resistor).

The analysis of the left part of the circuit is simplified by the very high input resistance of the FET. We can thus ignore the extremely small current into the gate and obtain the gate bias V_g from the voltage divider equation

$$V_g = \left(\frac{R_2}{R_1 + R_2}\right) V_{dd}. \tag{5.5}$$

The transistor is controlled, however, by the gate-source voltage V_{gs}, not by V_g. To obtain this, we start at the gate and complete the voltage loop to ground, giving $V_g - V_{gs} - I_d R_s = 0$. Note here that, because of the lack of any gate current, the current exiting the source is the same as that entering the drain, I_d. Solving for I_d gives

$$I_d = \frac{V_g - V_{gs}}{R_s}. \tag{5.6}$$

As usual, this equation is deceptively simple because I_d is a function of V_{gs} and V_{ds}. Since we intend to set our operating point in the saturation region, the dependence on V_{ds} is negligible and the dependence on V_{gs} is given by either Eq. (5.3) or the transfer curve. Using Eq. (5.3) we obtain

$$K(V_{gs} - V_t)^2 = \frac{V_g - V_{gs}}{R_s} \tag{5.7}$$

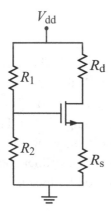

Figure 5.11 DC bias circuit.

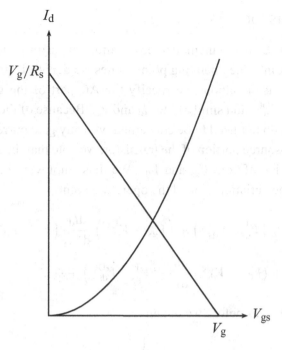

Figure 5.12 Graphical solution to Eq. (5.6).

which can be written

$$V_{gs}^2 + \left(\frac{1}{KR_s} - 2V_t\right)V_{gs} + \left(V_t^2 - \frac{V_g}{KR_s}\right) = 0. \tag{5.8}$$

This quadratic equation can then be solved for V_{gs}. This value is then used in Eq. (5.6) to obtain I_d. Alternatively, a graphical solution can be obtained by plotting the Eq. (5.6) load line and finding its intersection with the transfer curve, as shown in Fig. 5.12.

Turning to the right side of our bias circuit and applying the voltage loop law, we obtain $V_{dd} - I_d R_d - V_{ds} - I_d R_s = 0$ or

$$I_d = \frac{V_{dd} - V_{ds}}{R_s + R_d} \tag{5.9}$$

where V_{ds} is the voltage from the drain to the source of the transistor. Since we know I_d from our solution of Eq. (5.6), we can solve this for V_{ds} and complete our determination of the operating point. Alternatively, we can obtain a graphical solution by the load line method. The resulting plot will be similar to Fig. 5.10 except the y-intercept of the load line will be changed to $V_{dd}/(R_s + R_d)$.

5.4.2 AC equivalents for FETs

We now develop an equivalent circuit model for variations in the transistor behavior around the operating point. The operating point values we denote V_{gs}^{DC}, V_{ds}^{DC}, and I_d^{DC}. Using the lower-case notation, we specify the AC part of the gate-source voltage as $v_{gs} \equiv V_{gs} - V_{gs}^{DC}$, and similarly for i_d and v_{ds}. Because of the high input impedance of FETs we do not need to be concerned with any gate current.

Turning to the drain-source portion of the transistor, we note that, in general, the drain current is a function of both V_{gs} and V_{ds}. We thus employ a double Taylor expansion to express the variation around the operating point:

$$I_d(V_{gs}, V_{ds}) = I_d\left(V_{gs}^{DC}, V_{ds}^{DC}\right) + \left(V_{gs} - V_{gs}^{DC}\right)\frac{dI_d}{dV_{gs}}\left(V_{gs}^{DC}, V_{ds}^{DC}\right)$$

$$+ \left(V_{ds} - V_{ds}^{DC}\right)\frac{dI_d}{dV_{ds}}\left(V_{gs}^{DC}, V_{ds}^{DC}\right) + \cdots. \tag{5.10}$$

Introducing AC quantities as before we obtain

$$i_d = g_m v_{gs} + \frac{1}{r_{out}}v_{ds} \tag{5.11}$$

where

$$r_{out} \equiv \left[\frac{dI_d}{dV_{ds}}\left(V_{ds}^{DC}, V_{ds}^{DC}\right)\right]^{-1} \tag{5.12}$$

is the output (or drain) resistance and

$$g_m \equiv \frac{dI_d}{dV_{gs}}\left(V_{ds}^{DC}, V_{ds}^{DC}\right) \tag{5.13}$$

is the *transconductance*. The units of this latter quantity are variously named *amps per volt* (abbreviated A/V), *mhos* (℧), or *siemens* (S). Values of g_m vary from a few mA/V to values in the range of a few A/V. The transconductance is sometimes referred to as the *forward transfer conductance/admittance* and the symbol y_{fs} is used. Similarly, the inverse of r_{out} is sometimes called the *small signal admittance*, y_{os}.

Since our operating point is chosen to be in the saturation region, the drain current in Eqs. (5.12) and (5.13) is the saturation value of that current, $I_d = I_{d(sat)}$. We can then use Eq. (5.3) to obtain $g_m = 2K(V_{gs} - V_t)$.

The observant reader will also note that the I–V characteristics given in Figs. 5.2 and 5.10 are flat in the saturation region and, thus, $r_{out} = \infty$. In real FETs, there is a

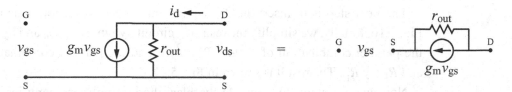

Figure 5.13 Completed AC equivalent for an FET.

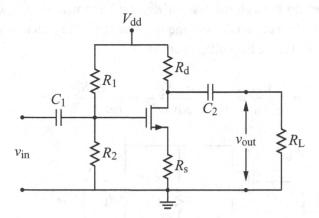

Figure 5.14 The common-source amplifier.

slight positive slope to these lines due to an effect called *channel length modulation*. The interested reader is referred to the references for a discussion of this effect. In the interests of simplicity, we will ignore this effect and take $r_{out} = \infty$.

The circuit equivalent of Eq. (5.11) is shown in Fig. 5.13. The drain current i_d is the sum of a current source with value set by the gate-source voltage and the transconductance and the current through the resistor r_{out} which is given by v_{ds}/r_{out}.

5.4.3 FET common-source amplifier

We now apply our AC equivalent to the analysis of an FET amplifier, shown in Fig. 5.14. Except for the substitution of an FET (here an e-mosfet) for the BJT, the configuration is the same as for the common-emitter amplifier studied in Chapter 4. The analysis of this common-source amplifier is also similar, but must take into account the differences in the AC model. Specifically, we will write the quantities v_{in}, v_{out}, i_{in}, and i_{out} in terms of v_{gs} instead of i_b.

Following the three step procedure of Chapter 4, we begin by treating the input and output capacitors as shorts and the DC power supply as ground. Recall that this approximation is valid when the capacitive impedances are negligible. The resulting circuit is shown in Fig. 5.15.

The next step is to insert the AC equivalent for the transistor. This is done in Fig. 5.16. Finally, we simplify the resulting circuit by ignoring r_{out} and by replacing the parallel combination of R_1 and R_2 by R_G and the parallel combination of R_d and R_L by R'_L. The result is shown in Fig. 5.17.

Now we are ready to calculate the black body amplifier quantities a, g, Z_{in}, and Z_{out} by writing v_{in}, v_{out}, i_{in}, and i_{out} in terms of v_{gs}. Since the current source current $g_m v_{gs}$ must go through the resistor R'_L, we have $v_{out} = -g_m v_{gs} R'_L$. The input voltage v_{in} is the gate-source voltage v_{gs} plus the voltage across R_s, giving $v_{in} = v_{gs} + g_m v_{gs} R_s$. Hence the voltage gain is

$$a = \frac{v_{out}}{v_{in}} = \frac{-g_m v_{gs} R'_L}{v_{gs} + g_m v_{gs} R_s} = \frac{-g_m R'_L}{1 + g_m R_s}. \tag{5.14}$$

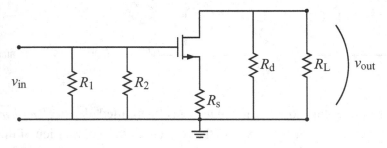

Figure 5.15 The common-source amplifier after applying step 1.

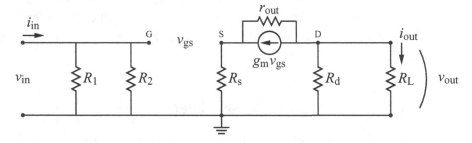

Figure 5.16 The common-source amplifier after inserting the transistor model.

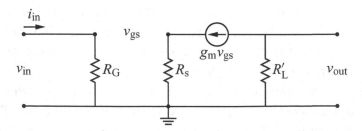

Figure 5.17 The common-source amplifier after simplifying.

Note that we have written both top and bottom of the quotient in terms of v_{gs} so that it would cancel out and leave us with an expression in terms of the circuit parameters. As usual, the minus sign in the answer indicates that the signal is inverted.

Next we compute the current gain and thus need expressions for i_{out} and i_{in}. Recalling that i_{out} is the current through the load resistor R_L, we have simply $i_{out} = v_{out}/R_L$. Computing i_{in} is particularly simple since the gate of the FET draws no current: $i_{in} = v_{in}/R_G$. Thus

$$g = \frac{i_{out}}{i_{in}} = \frac{v_{out}/R_L}{v_{in}/R_G} = \frac{-g_m}{1 + g_m R_s}\left(\frac{R_d R_G}{R_d + R_L}\right) \tag{5.15}$$

where we have substituted the expressions for v_{out} and v_{in} obtained in calculating a.

The computation of the input and output impedances is straightforward since we already have v_{in}, v_{out}, i_{in}, and i_{out}. Thus

$$Z_{in} = \frac{v_{in}}{i_{in}} = \frac{v_{in}}{v_{in}/R_G} = R_G \tag{5.16}$$

and

$$Z_{out} = \frac{v_{out}(R_L = \infty)}{i_{out}(R_L = 0)} = \frac{-g_m v_{gs} R_d}{-g_m v_{gs}} = R_d. \tag{5.17}$$

As with the bipolar transistor amplifiers, our results depend only on the known circuit parameters, so we can, in principle, adjust the values to match our requirements. Of particular interest is the result $Z_{in} = R_G = R_1 \| R_2$. We can thus make the input impedance of the amplifier as high as we like by making resistors R_1 and R_2 large. There is a limit, however, to this technique. While the FET gate impedance is very large, the gate does draw a small DC current. If we make the gate bias resistors R_1 and R_2 too large, this current will produce a large enough voltage drop across the resistors to upset the DC bias. While we could account for this effect in our circuit model, it is usually good enough to ignore the gate current but keep the values of R_1 and R_2 well below the gate input impedance.

5.4.4 FET common-drain amplifier

Our second example of an FET amplifier is the common-drain amplifier (also called the *source-follower*), shown in Fig. 5.18. In analogy with the BJT common-collector amplifier, the output is taken from the source of the transistor and the drain resistor is omitted. Application of the usual analysis steps leads to Fig. 5.19 where

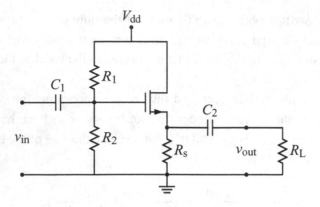

Figure 5.18 The common-drain amplifier.

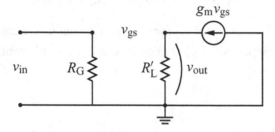

Figure 5.19 The common-drain amplifier after simplifying.

R'_L is the parallel combination of R_s and R_L; details are left as an exercise. We can now calculate the voltage gain a as before. Since the current source current $g_m v_{gs}$ must go through the resistor R'_L, we have $v_{out} = g_m v_{gs} R'_L$. The input voltage v_{in} is the gate-source voltage v_{gs} plus the voltage across R'_L, giving $v_{in} = v_{gs} + g_m v_{gs} R'_L$. Hence the voltage gain is

$$a = \frac{v_{out}}{v_{in}} = \frac{g_m v_{gs} R'_L}{v_{gs} + g_m v_{gs} R'_L} = \frac{g_m R'_L}{1 + g_m R'_L}. \tag{5.18}$$

Note that, as with the common-collector amplifier, this voltage gain is always less than one.

Derivation of the remaining the black body amplifier quantities g, Z_{in}, and Z_{out} is left as an exercise. The results are

$$g = \frac{g_m}{1 + g_m R'_L} \left(\frac{R_s R_G}{R_s + R_L} \right), \tag{5.19}$$

$$Z_{in} = R_G, \tag{5.20}$$

and

$$Z_{out} = R_s. \tag{5.21}$$

5.4.5 FET common-gate amplifier

Our final example of an FET amplifier is the common-gate amplifier shown in Fig. 5.20. This is the FET analog of the BJT common-base amplifier. For this amplifier we leave the entire analysis as an exercise. The results are:

$$a = g_m R_L',$$ (5.22)

$$g = \frac{g_m}{1 + g_m R_s} \left(\frac{R_s R_d}{R_d + R_L} \right),$$ (5.23)

$$Z_{in} = \frac{R_s}{1 + g_m R_s},$$ (5.24)

and

$$Z_{out} = R_d.$$ (5.25)

Here R_L' is the parallel combination of R_d and R_L.

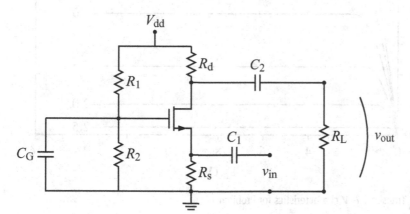

Figure 5.20 The common-gate amplifier.

EXERCISES

1. Consider the transistor characteristics of Fig. 5.21. (a) Are these the characteristics of a JFET, d-mosfet, or e-mosfet? (b) Make a table of the given V_{gs} values and the corresponding $I_{d(sat)}$ values. Also include a column in the table giving $\sqrt{I_{d(sat)}}$. (c) Make a plot of $\sqrt{I_{d(sat)}}$ versus V_{gs}. Find the slope and y-intercept of the plot and use these to determine values for K and V_t in the model equation Eq. (5.3).

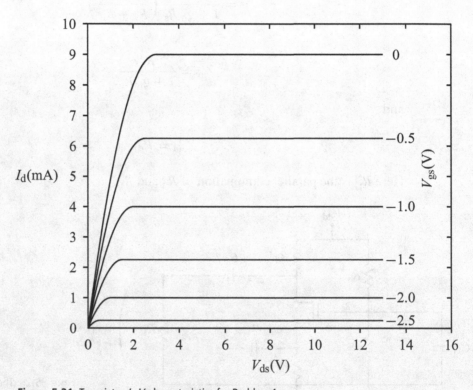

Figure 5.21 Transistor I–V characteristics for Problem 1.

2. Determine the operating point (V_{gs}, I_d, and V_{ds}) of a universal transistor DC bias circuit when $V_{dd} = 15$ V, $R_1 = 1$ MΩ, $R_2 = 100$ kΩ, $R_d = 3000$ Ω, and $R_s = 1000$ Ω. Use the values of K and V_t determined in Problem 1.

3. For the operating point of Problem 2, determine the AC model parameters g_m and r_{out}.

4. If the circuit of Problem 2 is configured as a common-source amplifier, calculate the resulting voltage and current gain. Assume a load resistor of 10 kΩ and use the AC model parameters determined in Problem 3.

5. Derive the expressions for the current gain g, the input impedance Z_{in}, and the output impedance Z_{out} for the common-drain amplifier.
6. Derive the expressions for the voltage gain a, the current gain g, the input impedance Z_{in}, and the output impedance Z_{out} for the common-gate amplifier.

FURTHER READING

Richard C. Jaeger and Travis N. Blalock, *Microelectronic Circuit Design*, 3rd edition (New York: McGraw-Hill, 2008).

Paul Horowitz and Winfield Hill, *The Art of Electronics*, 2nd edition (New York: Cambridge University Press, 1989).

Donald A. Neamen, *Microelectronics: Circuit Analysis and Design*, 3rd edition (New York: McGraw-Hill, 2007).

John E. Uffenbeck, *Introduction to Electronics, Devices and Circuits* (Englewood Cliffs, NJ: Prentice-Hall, 1982).

6 Operational amplifiers

6.1 Introduction

We now turn to an examination of the properties and uses of the *operational amplifier* or *op-amp*. A detailed analysis of this multi-stage amplifier circuit is beyond the scope of this text, so we will treat it as a black box device as we did earlier with the voltage regulator. Thus, to use the device, we need only learn and apply some simple rules and, later, the real-world limitations of the device.

In current usage, the operational amplifier is usually packaged as an integrated circuit with multiple leads or pins. While there are hundreds of different op-amps with different specifications, they all follow the same usage rules. To be specific, we will focus on a "classic" version: the 741 op-amp.

The circuit symbol for the op-amp is shown in Fig. 6.1. There are inputs for two power supply voltages (one positive and one negative relative to ground, labeled V_{cc}^+ and V_{cc}^-, respectively). There are also two signal inputs: the *inverting input*, labeled with a minus sign, and the *non-inverting input*, labeled with a plus sign. Happily, there is only one output.

As we know, voltages are always between two points, but our description of the op-amp inputs seems to refer to voltages at one point, the various input pins. It is thus important to note that all of the voltages for the op-amp are referenced to ground (i.e., the second point is ground). While it is common for writers discussing op-amp circuits to refer to the voltage at some point, one should keep in mind that they are really talking about the voltage between this point and ground. Also, the power supply connections shown in Fig. 6.1 are often omitted from circuit diagrams for simplicity, and it is easy for the novice building such a circuit to forget these connections. Of course, the circuit will not work without them.

The basic operation of the op-amp can be simply stated. The output voltage is proportional to the difference between the inverting and non-inverting input voltages:

$$V_{out} = A_{OL} \left(V_{in}^+ - V_{in}^- \right) \tag{6.1}$$

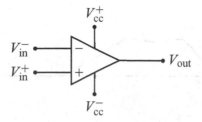

Figure 6.1 Schematic symbol for the operational amplifier.

where A_{OL} is the *open-loop voltage gain*. Note that the plus and minus signs on the two input voltages are simply labels denoting which input the voltage is applied to; they do *not* specify the polarity of the input voltages.

Typically, A_{OL} is very large (for the 741 it is 200 000). One might then imagine huge output voltages, but Eq. (6.1) is subject to limitations. The output voltage V_{out} can only be within a range set by two *saturation voltages*:

$$V_{sat}^- \leq V_{out} \leq V_{sat}^+ \tag{6.2}$$

where $V_{sat}^+ \approx V_{cc}^+ - 1\,V$ (a little below the positive power supply voltage) and $V_{sat}^- \approx V_{cc}^- + 1\,V$ (a little above the negative power supply voltage). The output current is also restricted: for the 741 op-amp it must be less than 25 mA. Finally, we note that the input impedance of the two inputs is very high, so very little current flows into these inputs.

The restrictions imposed by Eq. (6.2) along with the large value of A_{OL} mean that any small difference between the op-amp inputs will cause the output to saturate. For example, if we use the 741 op-amp with $\pm15\,V$ power supplies, it takes only a 70 μV difference between the inputs for the output to reach its limit. This is the basis for the non-linear applications of the op-amp.

6.2 Non-linear applications I

Non-linear applications of the op-amp use the device as a *comparator*. It compares the voltage at the two inputs and gives a positive or negative output depending on which input is larger. Such an application is shown in Fig. 6.2. The input is an arbitrary signal that wanders around the level applied to the inverting input (here, 1 V). When the input signal is above 1 V, the output swings to its positive saturation limit.[1] When the input falls below 1 V, the output falls to its negative saturation limit. The state of the output thus tells us the relative size of the input

---

[1] Strictly speaking, it must be 70 μV above 1 V, but this is so small as to be negligible.

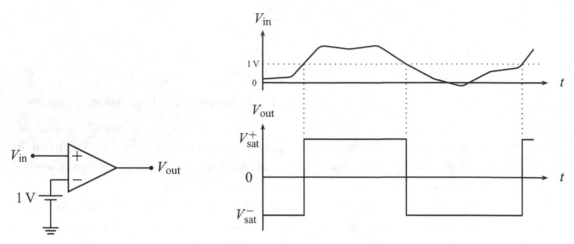

Figure 6.2 Using the op-amp as a comparator.

and the voltage applied to the inverting input. The relationship between the input and output signals is clearly not proportional (i.e., the operation is non-linear).

6.3 Linear applications

We will see additional examples of non-linear applications later, but now we turn to the more common linear applications. As we shall see, these circuits are characterized by a single feedback connection from the output to the inverting input of the op-amp. The circuits can be analyzed by applying two approximate rules called the golden rules of ideal linear op-amp operation.

1. The output will do whatever is necessary to make the voltage difference between the inputs zero.
2. No current flows into the inputs.

The application of these rules is best illustrated by working several examples. The first of these is shown in Fig. 6.3. An input voltage is applied to the inverting input through resistor R_1 and a feedback resistor R_f is connected between this input and the output. Our method of applying rule 1 is to take the voltage at the two inputs to be equal and then determine the output voltage required to achieve this. Since the non-inverting (+) input is connected to ground, we take the voltage at

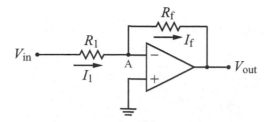

Figure 6.3 Simple inverting amplifier.

point A also to be zero. Ohm's Law tells us that $V_{in} - V_A = I_1 R_1$ so

$$I_1 = \frac{V_{in} - V_A}{R_1} = \frac{V_{in}}{R_1}. \tag{6.3}$$

Next, we apply KVL starting at point A:

$$V_A - I_f R_f - V_{out} = 0. \tag{6.4}$$

But, $V_A = 0$ (by rule 1) and $I_f = I_1$ (by rule 2). Applying these and the result of Eq. (6.3) to Eq. (6.4) we obtain

$$V_{out} = V_A - I_f R_f = -I_1 R_f = -\frac{R_f}{R_1} V_{in}. \tag{6.5}$$

Thus the output is proportional to the input, with a proportionality constant $-R_f/R_1$ set by the two resistors. The minus sign means the signal is inverted.

Recalling our black box model for an amplifier, the input resistance is just the effective resistance between the input terminals of the amplifier. For this circuit, the input terminals are the one marked V_{in} and ground. Since point A is at ground potential, the input resistance is simply R_1. The output resistance is not so simply found, but the op-amp is designed to have a very low output impedance.

We thus have a nice voltage amplifier with easily selectable input resistance and voltage gain set by two resistor values. If we contrast this with the relative difficulty of designing a transistor amplifier, it is easy to see why the op-amp is such a popular component.

Our next application is shown in Fig. 6.4. In this case, the input voltage is applied directly to the (+) input.[2] By rule 1, the voltage at point A is equal to the input voltage, $V_A = V_{in}$. By Ohm's Law, the current through resistor R_1 is given by $I_1 = V_A/R_1$. Rule 2 tells us that $I_2 = I_1$, since no current flows into (or out of) the (−) input. Applying KVL starting at point A gives

$$V_A + I_2 R_2 - V_{out} = 0. \tag{6.6}$$

---

[2] Note that, for convenience, we have flipped the position of the (+) and (−) inputs on the op-amp circuit symbol.

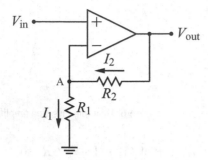

Figure 6.4 The non-inverting amplifier.

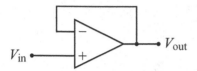

V_{out}

Figure 6.5 The buffer or voltage follower.

Solving for V_{out} and applying our other results yields

$$V_{\text{out}} = \left(1 + \frac{R_2}{R_1}\right) V_{\text{in}}. \tag{6.7}$$

Our analysis shows that this circuit is a *non-inverting amplifier* with a gain set by the ratio R_2/R_1. Interestingly, as a consequence of rule 2, the input impedance for this amplifier is infinite.[3]

Figure 6.5 shows the *buffer circuit* (also called a *voltage follower*). The analysis here is particularly simple since V_{out} is directly connected to the $(-)$ input, and, by rule 1, the voltage at the $(-)$ input is equal to the voltage at the $(+)$ input. Thus, for this circuit, the output is the same as the input, $V_{\text{out}} = V_{\text{in}}$. This does not seem particularly useful until one considers the input and output impedances. As with the non-inverting amplifier, the input impedance is infinite, and, as with all op-amp circuits, the output impedance is low. This circuit can thus be used between a high impedance voltage source and a low impedance load to alleviate loading due to the voltage divider effect.[4]

An *adder circuit* is shown in Fig. 6.6. The version shown here has three inputs, but any number is possible. As before, rule 1 tells us that $V_A = 0$. Ohm's Law then gives us the current through each of the input resistors: $I_1 = V_1/R_1$, $I_2 = V_2/R_2$, and $I_3 = V_3/R_3$. These three currents add (by KCL), and, since none of the current

---

[3] Of course, in the real world, there are no infinities, but the input impedance will be very large.
[4] Recall that this same role is played by the common-collector transistor amplifier.

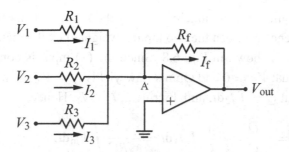

Figure 6.6 The adder circuit.

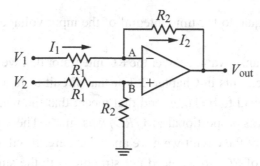

Figure 6.7 The differential amplifier.

flows into the $(-)$ input, $I_f = I_1 + I_2 + I_3$. Finally, KVL gives $V_A - I_f R_f - V_{out} = 0$. Putting this all together then yields

$$V_{out} = -\left(\frac{R_f}{R_1} V_1 + \frac{R_f}{R_2} V_2 + \frac{R_f}{R_3} V_3 \right). \tag{6.8}$$

Our circuit thus gives us a *weighted sum* of the input voltages with the weighting determined by the resistor values. If we want a simple unweighted sum, we simply choose all the resistor values to be the same. Note that our analysis is easily extended to include any number of inputs: each input will add a term to Eq. (6.8) of the form $\frac{R_f}{R_i} V_i$ where i is the input number.

Op-amps can be used to perform other mathematical functions as well. Figure 6.7 shows the *differential amplifier* circuit which can be used to subtract two voltages. Because of rule 2, no current flows into the $(+)$ input, so the resistors R_1 and R_2 attached to that input form a voltage divider with

$$V_B = \left(\frac{R_2}{R_1 + R_2} \right) V_2. \tag{6.9}$$

By Ohm's Law, $I_1 = (V_1 - V_A)/R_1$ and, by KVL, $V_{out} = V_A - I_2 R_2$. By rule 1, $V_A = V_B$, and by rule 2, $I_2 = I_1$. Putting this all together yields

$$V_{out} = V_A - I_2 R_2 = \left(\frac{R_2}{R_1 + R_2} \right) V_2 - \left(\frac{V_1}{R_1} - \frac{R_2}{R_1 + R_2} \frac{V_2}{R_1} \right) R_2 = \frac{R_2}{R_1}(V_2 - V_1) \tag{6.10}$$

where this last step requires a little algebra (left to the reader). Thus the output voltage gives the difference between the two inputs, weighted by the factor R_2/R_1.

An *integrator circuit* is shown in Fig. 6.8. Since the (+) input is connected to ground, rule 1 tells us that $V_A = 0$ and Ohm's Law yields $I_1 = V_{in}/R$. KVL gives $V_A - \frac{Q}{C} - V_{out} = 0$. But $Q = \int I_2 dt$, and, by rule 2, $I_2 = I_1$. Hence,

$$V_{out} = -\frac{Q}{C} = -\frac{1}{C} \int I_1 dt = -\frac{1}{RC} \int V_{in} dt. \tag{6.11}$$

In this case, the output is equal to the time integral of the input voltage weighted by the factor $-1/RC$.

This circuit has an important advantage over the RC integrator that we examined back when we studied RC circuits in Chapter 2. For that circuit to function as an integrator, the value of RC had to be large, and that meant that the magnitude of the output voltage (which was proportional to $1/RC$) was small. There is no such restriction on the operation of the circuit we have analyzed here. It will operate as an integrator for any value of RC, so we need not struggle with the small output signal levels characteristic of the RC integrator.

Finally, we consider the *differentiator circuit* of Fig. 6.9. As before, $V_A = 0$, so $V_{in} = Q/C$. Taking the time derivative of this gives

$$\frac{dV_{in}}{dt} = \frac{1}{C}\frac{dQ}{dt} = \frac{1}{C}I_1. \tag{6.12}$$

But KVL gives $V_{out} = -I_2 R$ and, by rule 2, $I_2 = I_1$. Thus

$$V_{out} = -RC\frac{dV_{in}}{dt}. \tag{6.13}$$

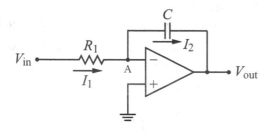

Figure 6.8 The op-amp integrator.

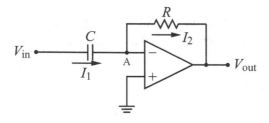

Figure 6.9 The op-amp differentiator.

The output is the time derivative of the input, weighted by the factor $-RC$. Again, note that our derivation places no restriction on the value of RC.

6.4 Practical considerations for real op-amps

Up to this point we have assumed our op-amp circuits were governed by Eq. (6.1) and Eq. (6.2) for non-linear circuits and by the golden rules for linear circuits. In practice, there are a few complications, and these need to be understood to use operational amplifiers successfully.

6.4.1 Bias currents

Golden rule 2 says that no current flows into the inputs of the op-amp. Actually, a small DC current *must* flow into each input. These *bias currents* are denoted I_B^+ and I_B^- for the (+) and (−) inputs respectively. For the 741 op-amp, the bias currents are specified as less than 500 nA. Because they are so small, we can usually ignore them; that is why golden rule 2 works. Under certain circumstances, however, they can cause large effects.

Consider the circuit shown in Fig. 6.10. This is just the non-inverting amplifier of Fig. 6.4 with a capacitor placed on the input (perhaps to filter out the DC component of the input signal). This seemingly innocent addition, however, will keep the circuit from working at all: the output will be saturated at V_{sat}^+ and unresponsive to the input. The reason is that the capacitor has blocked the DC bias current for the (+) input. Since this current is not present, the op-amp will not work.

Another consequence of non-zero bias currents is illustrated in Fig. 6.11. This, again, is the non-inverting amplifier, but now we have grounded the input, perhaps to test the circuit; if the input is zero, we expect from Eq. (6.7) that the output will be zero. Since the (+) input is grounded, point A will also be zero volts (by rule 1),

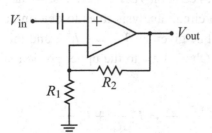

Figure 6.10 An AC coupled non-inverting amp?

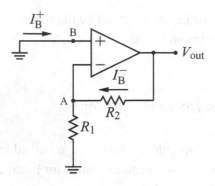

Figure 6.11 Bias currents in the non-inverting amp.

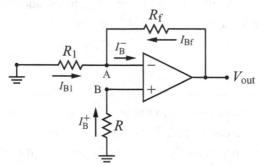

Figure 6.12 General analysis of bias current effects.

and no current will flow through resistor R_1. The bias current I_B^- must therefore flow through R_2, making the output voltage

$$V_{\text{out}} = I_B^- R_2 = (500 \text{ nA})(1 \text{ M}\Omega) = 0.5 \text{ V} \tag{6.14}$$

where we have used the maximum 741 bias current and a large feedback resistor value $R_2 = 1$ MΩ for the calculation. This is certainly not the expected zero volts output and is totally incomprehensible unless one is aware of the bias currents. One lesson to be drawn from this example is to avoid large feedback resistors since they enhance the effect of the bias current.

While bias currents are always present, there is a clever way to minimize their effects. Consider the circuit shown in Fig. 6.12. As we will see, this can represent a slightly modified version of the inverting or non-inverting amplifiers, or the buffer circuit, but since we are interested in bias current effects, we have grounded the inputs. We now apply the usual circuit analysis rules to determine the output voltage. The voltage at point B will be given by $V_B = -I_B^+ R$ and this, by rule 1, is equal to V_A. Applying KCL and Ohm's Law to the upper portion of the circuit gives

$$I_B^- = I_{B1} + I_{Bf} = -\frac{V_A}{R_1} + \frac{V_{\text{out}} - V_A}{R_f} = I_B^+ \frac{R}{R_1} + I_B^+ \frac{R}{R_f} + \frac{V_{\text{out}}}{R_f}. \tag{6.15}$$

Solving this for V_{out} then yields

$$V_{\text{out}} = I_{\text{B}}^- R_{\text{f}} - I_{\text{B}}^+ R \left(1 + \frac{R_{\text{f}}}{R_1}\right). \tag{6.16}$$

If we now choose

$$R = \frac{R_1 R_{\text{f}}}{R_1 + R_{\text{f}}} \tag{6.17}$$

Eq. (6.16) reduces to

$$V_{\text{out}} = R_{\text{f}} \left(I_{\text{B}}^- - I_{\text{B}}^+\right). \tag{6.18}$$

The resistor choice of Eq. (6.17) thus gives us an output that depends on the *difference* between the bias currents. This should be compared to the case in Eq. (6.14) where the output depends directly on I_{B}^-. In a perfectly symmetric op-amp, the two bias currents would be the same and we would have totally mitigated the effect of the bias currents. For real op-amps, the maximum difference is specified as the *input offset current* $I_{\text{os}} \equiv I_{\text{B}}^- - I_{\text{B}}^+$. For the 741 op-amp, $I_{\text{os}} < I_{\text{B}}^{\pm}/4$. If this is still too large to tolerate, a higher quality op-amp will often sport better specifications for the bias currents (e.g., $I_{\text{B}}^{\pm} < 0.05$ nA for the CA3130 op-amp).

Application of this mitigation technique to some of our previous circuits is shown in Figs. 6.13, 6.14, and 6.15. For the first two of these, the resistor R with value given by Eq. (6.17) is added to the (+) input of the op-amp. The reader can verify that both of the resulting circuits match the circuit of Fig. 6.12 if the inputs are grounded. Note that these additional resistors make no sense at all unless we know about bias currents: since there is no current into the (+) input there is no voltage drop across R, so including it makes no difference.

The modifications to the buffer circuit in Fig. 6.15 can be understood as a special case of Fig. 6.12 where $R_1 \to \infty$. In this case, Eq. (6.17) yields $R = R_{\text{f}}$, so the mitigation effect is obtained for any resistor value, as long as the two are equal.

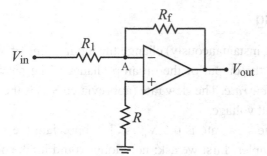

Figure 6.13 Applying the bias current mitigation technique to the inverting amplifier.

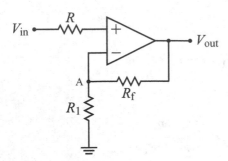

Figure 6.14 Applying the bias current mitigation technique to the non-inverting amplifier.

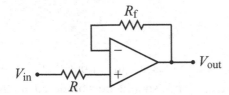

Figure 6.15 Applying the bias current mitigation technique to the buffer.

6.4.2 Input offset voltage

Asymmetries between the internal circuits driven by the (+) and (−) inputs can also lead to non-zero output voltages. The amount of voltage necessary at the input terminals to return the output to zero is called the *input offset voltage*, V_{io}. This offset voltage acts like a small voltage source ($V_{io} < 5$ mV for the 741) in series with one of the inputs. Thus the simple experiment of taking an op-amp and grounding both inputs would not produce a zero output, as expected from Eq. (6.1), but a saturated output (remember it only takes about 70 μV difference in the inputs to saturate the output). The solution in this case is built into the op-amp circuit. Most op-amps have additional inputs for *external balancing* or *nulling*. One simply connects the ends of a potentiometer (typically 10 kΩ) to these inputs and connects the slider of the potentiometer to V_{cc}^-. The potentiometer is then adjusted to reduce to output to zero.

6.4.3 Slew rate limiting

Suppose we suddenly (i.e., instantaneously) change the input of one of our op-amp amplifiers. How fast will the output of the op-amp change? The answer to this question depends on the *slew rate*. The slew rate (abbreviated SR) is the maximum rate of change of the output voltage.

For the 741 op-amp, the slew rate is 0.5 V/μs. To appreciate the meaning of this, let's look at two examples. First we ask: how long would it take our op-amp

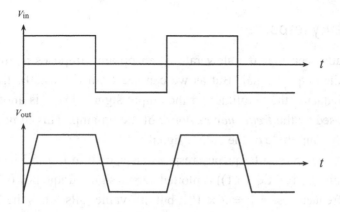

Figure 6.16 Distortion of a square wave due to slew rate limiting.

to switch from its minimum voltage V_{sat}^- to its maximum voltage V_{sat}^+? Assuming ± 15 V power supplies, $V_{sat}^{\pm} = \pm 14$ V, and we have

$$t = \frac{\Delta V}{SR} = \frac{28 \text{ V}}{0.5 \text{ V/}\mu s} = 56 \text{ }\mu s. \tag{6.19}$$

This is a fairly slow response. The result can be a distorted output waveform, as is shown in Fig. 6.16. The output tries to respond to the changed input (in this case a square wave) as fast as it can, but is limited by the slew rate.

As a second example, imagine that the output is a sine wave of the form $A \sin \omega t$. The rate of change of this output is then $dV/dt = \omega A \cos \omega t$. To avoid slew rate limiting, we want the maximum rate of change to be less than the slew rate, or $\omega A < $ SR. If we suppose that $A = 10$ V, then, for the 741, the highest frequency we can handle is

$$f_{max} = \frac{SR}{2\pi A} = \frac{0.5 \text{ V/}\mu s}{2\pi \cdot 10 \text{ V}} \approx 8 \text{ kHz}. \tag{6.20}$$

Thus the frequency that can be amplified without distortion is rather limited. Note that this limit on the frequency decreases as the amplitude of the output signal increases. The value of f_{max} when $A = V_{sat}$ is called the *full power bandwidth*.

The only way to avoid the problem of slew rate limiting is to obtain an op-amp with better specifications. For example, the CA3130 op-amp has a slew rate of 10 V/μs, twenty times better than the 741. There is, however, a trade-off between slew rate and stability. An op-amp with high slew rate will often also be susceptible to instability – an amplifier circuit built with such an op-amp may start to oscillate spontaneously.

6.4.4 Frequency response

We have already seen that the slew rate of an op-amp imposes a limit on the frequency of the output signal. But as we can see from Eq. (6.20), this can be alleviated by reducing the amplitude of the output signal. There is another limit, however, imposed by the *frequency response* of the op-amp. This limit is present regardless of the amplitude of the output signal.

A representative op-amp frequency response is shown in Fig. 6.17. The open-loop voltage gain A_{OL} (cf. Eq. (6.1)) is plotted versus signal frequency on a log-log scale. At low frequencies, $A_{OL} = 2 \times 10^5$, but this value rolls off as the frequency increases. This roll-off is deliberately designed into the op-amp to avoid oscillation at high frequencies. Since the impedance of stray capacitances decreases with frequency, it is easier for high frequency signals to couple between the output and the input of the op-amp, thus producing unintentional positive feedback and, hence, oscillation. The roll-off counteracts this by reducing the amount these stray signals are amplified.

This frequency response also affects the gain of our linear circuits (e.g., the inverting and non-inverting amplifiers of Figs. 6.3 and 6.4). This gain, often called the *closed loop gain* A_{CL}, is shown by the horizontal dotted line in Fig. 6.17 for the particular case where $A_{CL} = 10^3$. The gain remains at this value as the signal frequency increases until it intersects with the open loop gain curve, after which it, too, rolls off.

The frequency range over which A_{CL} is constant is called the *closed loop bandwidth*, B_{CL}. For the portion of the curve where A_{OL} is decreasing with frequency, the product $A_{CL}B_{CL}$ is a constant, called either the *gain bandwidth product* or the *open*

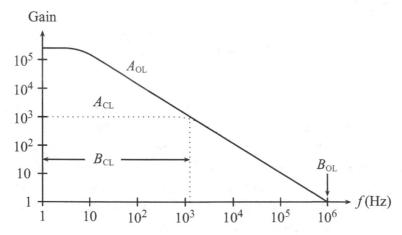

Figure 6.17 Frequency response of the 741 op-amp.

loop bandwidth B_{OL}. This latter quantity is also defined as the frequency where the gain falls to one. For the 741, $B_{OL} = 1$ MHz (cf. Fig. 6.17). Thus

$$B_{CL} = \frac{B_{OL}}{A_{CL}}. \tag{6.21}$$

A high gain amplifier will therefore have a smaller bandwidth and vice versa. If you want an amplifier with a large bandwidth, you either have to decrease the amplifier closed-loop gain or buy an amplifier with a larger value for B_{OL}.

6.5 Non-linear applications II

As a lead-in to our next chapter and another example of a non-linear application, consider the circuit in Fig. 6.18, the *op-amp astable multivibrator*. At first glance, this may appear to be a linear circuit because it has feedback between the output and the $(-)$ input. However, it also has feedback between the output and the $(+)$ input. Linear circuits only have the negative feedback connection.

Because of all the feedback connections, it is difficult to know where to begin in our analysis of this circuit. We start by assuming the the capacitor is initially uncharged (as it would be if we had just constructed the circuit). The voltage at point A (the inverting input of the op-amp) is thus zero. If we suppose that the output is zero, then the voltage at point B (the non-inverting input) is also zero, and the circuit will remain in this state since Eq. (6.1) is satisfied. This, of course, would not be a very interesting circuit.

The inevitable circuit noise, however, will cause the output voltage to fluctuate slightly. Suppose the output voltage becomes slightly positive.[5] The voltage divider formed by resistors R_1 and R_2 will then produce a positive voltage at point B. Since point A is initially zero, this positive difference between the inputs produces a much larger positive voltage at the output (cf. Eq. (6.1)). The feedback then makes the voltage at the $(+)$ input larger and the output quickly becomes saturated: $V_{out} \rightarrow V_{sat}^+$. The voltage at point B is now also constant with value $V_{sat}^+ R_2/(R_1 + R_2)$.

The capacitor C starts to charge up through resistor R_f with time constant $R_f C$, driven by the output voltage. We know from our studies of RC circuits that, if we let the charging process go on, the voltage across the capacitor would eventually reach the output voltage (which at this point is V_{sat}^+). But this will never happen, because as soon as the voltage at point A exceeds the voltage at point B (which is a fraction

---

[5] The analysis is not crucially dependent on this assumption; the circuit will also start to oscillate if we assume the fluctuation is negative.

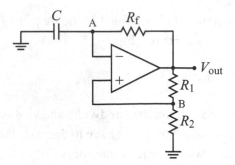

Figure 6.18 The op-amp astable multivibrator.

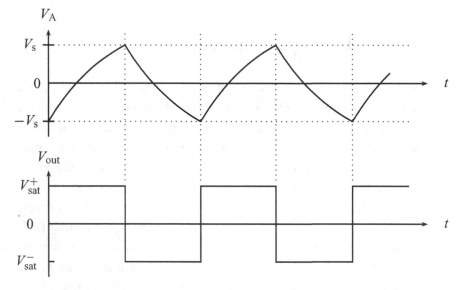

Figure 6.19 Resulting waveforms for the astable multivibrator.

of V_{sat}^+), the voltage difference between the inputs will be negative and the output will be driven to its negative saturation value, $V_{\text{out}} \rightarrow V_{\text{sat}}^-$. This is reinforced by the fact that point B now becomes negative with value $V_{\text{sat}}^- R_2/(R_1 + R_2)$. The capacitor now charges toward V_{sat}^- until point A falls below point B, at which point the output switches to is positive saturation value. The process continues and produces a square wave output as shown in Fig. 6.19. We have also shown in this figure the voltage at point A to emphasize its role in setting the timing for the output switching. Here we have defined the switch voltage $\pm V_s \equiv \pm V_{\text{sat}}^+ R_2/(R_1 + R_2)$ and have assumed that $V_{\text{sat}}^- = -V_{\text{sat}}^+$, as is typically the case.

Let's quantify this analysis by deriving an expression for the period of the output waveform. From our qualitative discussion, it is clear that the period is set by the charging of the capacitor. The relevant portion of our circuit is shown in Fig. 6.20.

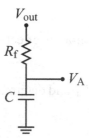

Figure 6.20 The RC circuit portion of the astable multivibrator.

Recall from Chapter 2, Eq. (2.22), that the general solution for the RC charging problem is given by

$$V_A = V_1 e^{-t/R_f C} + V_2 \tag{6.22}$$

where V_1 and V_2 are constants and we have included the fact that, in our problem, the voltage across the capacitor is V_A. The constants are determined by two limits: at $t = 0$, $V_A = -V_s$, and for $t \rightarrow \infty$, $V_A \rightarrow V_{sat}^+$ (although, in this problem, this point is never reached). Using the second limit in Eq. (6.22) gives us $V_2 = V_{sat}^+$. Using this result and the $t = 0$ limit, the reader can verify that

$$V_1 = -V_{sat}^+ \left[\left(\frac{R_2}{R_1 + R_2} \right) + 1 \right] \tag{6.23}$$

so our general solution becomes

$$V_A = -V_{sat}^+ \left[\left(\frac{R_2}{R_1 + R_2} \right) + 1 \right] e^{-t/R_f C} + V_{sat}^+. \tag{6.24}$$

Referring to Fig. 6.19, we want to know the time t_0 when $V_A = +V_s$ since this will give us the time when the output switches from positive to negative. Plugging into Eq. (6.24) gives

$$\left(\frac{R_2}{R_1 + R_2} \right) V_{sat}^+ = -V_{sat}^+ \left[\left(\frac{R_2}{R_1 + R_2} \right) + 1 \right] e^{-t_0/R_f C} + V_{sat}^+. \tag{6.25}$$

Solving this for t_0 yields (after some algebra)

$$t_0 = R_f C \ln \left(1 + \frac{2R_2}{R_1} \right). \tag{6.26}$$

Since the discharge portion of the capacitor signal is symmetric with the charging portion, the period of our square wave oscillator is just twice this value: $T = 2t_0$. Note that the period is set primarily by the product $R_f C$, as we might have expected. There is a weaker dependence on resistors R_1 and R_2, since these define V_s. Interestingly, the dependence on V_{sat}^+ has canceled out, so variations in the power supply will not affect the frequency of this oscillator.

EXERCISES

Note: for these exercises you can ignore non-ideal op-amp effects (e.g., bias currents).

1. For the circuit of Fig. 6.21, find V_{out} as a function of V_{in} and determine the input impedance of the circuit.

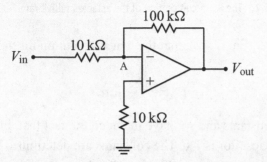

Figure 6.21 Circuit for Problem 1.

2. Derive an expression for the output voltage of the circuit in Fig. 6.22 in terms of the four input voltages. Simplify your result as much as possible.

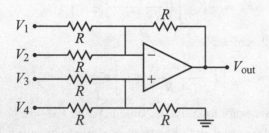

Figure 6.22 Circuit for Problem 2.

3. For the circuit in Fig. 6.23, find an expression for the current through the ammeter in terms of the input voltage. Assume the meter has zero internal resistance. How does the sign of the input voltage affect the answer?

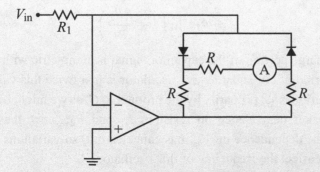

Figure 6.23 Circuit for Problem 3.

4. What is the maximum peak-to-peak output voltage for a 741 op-amp with ±15 V power supplies? How small a load resistance will this amplifier drive with its output at the maximum level?

5. Refer to Fig. 6.24 and determine the output voltage for each circuit. Assume $V_{\text{sat}}^{\pm} = \pm 10$ V.

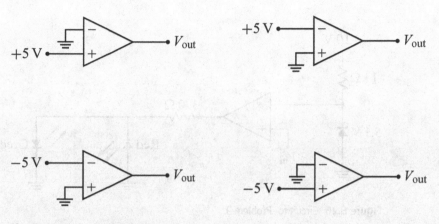

Figure 6.24 Circuits for Problem 5.

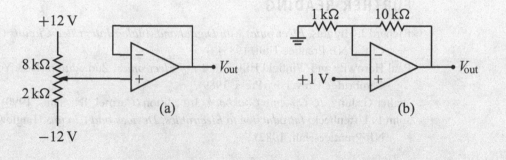

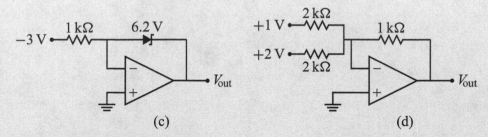

Figure 6.25 Circuits for Problem 8.

6. Draw the schematic diagram of an inverting amplifier with $R_i = 5$ kΩ and a voltage gain of -75.
7. If the circuit of the previous problem uses ±12 V power supplies, what input voltage will cause the output to saturate?
8. Determine the output voltage for each circuit in Fig. 6.25.
9. Explain the operation of the circuit shown in Fig. 6.26 as V_{in} is varied.

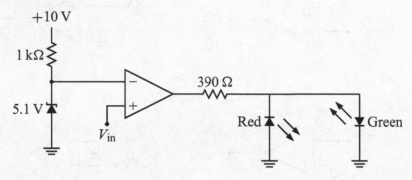

Figure 6.26 Circuit for Problem 9.

FURTHER READING

Richard J. Higgins, *Electronics with Digital and Analog Integrated Circuits* (Englewood Cliffs, NJ: Prentice-Hall, 1983).

Paul Horowitz and Winfield Hill, *The Art of Electronics*, 2nd edition (New York: Cambridge University Press, 1989).

Walter G. Jung, *IC Op-Amp Cookbook*, 3rd edition (Carmel, IN: Sams, 1990).

John E. Uffenbeck, *Introduction to Electronics, Devices and Circuits* (Englewood Cliffs, NJ: Prentice-Hall, 1982).

Oscillators

7.1 Introduction

The op-amp astable oscillator covered in Section 6.5 was our first example of an oscillator – a circuit that produces a periodic output signal without an input signal. These types of circuits have some kind of feedback mechanism that allows them to oscillate spontaneously. We can categorize oscillators into two broad groups: *relaxation oscillators* and *sinusoidal oscillators*. Each of these groups has common features. The relaxation oscillators are characterized by non-sinusoidal output waveforms, timing that is set by capacitor charging and discharging, and the non-linear operation of its active components. The analysis of relaxation oscillator circuits is done in the time domain (i.e., by determining the circuit response as a function of time). For example, our op-amp astable oscillator has a square wave output with a period set by the charging/discharging of capacitor C through resistor R_f, and the op-amp is operating non-linearly, switching back and forth between its saturation voltages. On the other hand, sinusoidal oscillators, as the name implies, have sinusoidal output waveforms and linear operation of the active components, and the analysis is done in the frequency domain (i.e., by considering how the circuit responds to different frequencies). We will now examine examples of each type of oscillator.

7.2 Relaxation oscillators

7.2.1 SCR sawtooth oscillator

Our first relaxation oscillator is shown in Fig. 7.1. It uses two components we have studied previously: the SCR and the bipolar transistor. To begin, recall that the SCR will remain in its off-state unless there is current flowing into the gate[1] and no

---

[1] Here we assume that the power supply voltage is less than the critical voltage for the case of zero gate current $V_{cc} < V_{crit0}$.

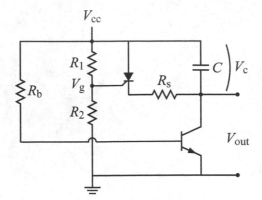

Figure 7.1 The SCR sawtooth oscillator.

gate current will flow unless the voltage at the gate V_g is greater than the voltage at the cathode. We can then construct a description of the circuit operation as follows.

1. The voltage at the gate of the SCR, V_g, is set by the voltage divider formed by R_1 and R_2 to the value

$$V_g = \left(\frac{R_2}{R_1 + R_2}\right) V_{cc}. \tag{7.1}$$

If we assume the capacitor is initially uncharged (as it would be if we had just turned on the circuit), then the voltage across the capacitor V_c is zero and $V_{out} = V_{cc}$. Thus $V_g < V_{out}$ and no gate current can flow, so the SCR does not turn on.

2. The transistor has a fixed base current

$$I_b = \frac{V_{cc} - 0.6}{R_b} \tag{7.2}$$

and thus will have a constant collector current $I_c = \beta I_b$. Since the SCR is off, this current must flow through the capacitor, thus charging it at a constant rate $dQ/dt = I_c$ and increasing V_c.

3. Eventually, V_c becomes large enough (and, thus, V_{out} small enough) that $V_g > V_{out}$. The SCR then turns on and discharges the capacitor through the resistor R_s. R_s is made small enough that this discharge happens quickly, but must be large enough to prevent peak currents from destroying the SCR.

4. As the capacitor discharges, V_c decreases and V_{out} increases. The gate current goes to zero and, as the current through the SCR drops below I_{crit}, the SCR shuts off.[2]

---

[2] This assumes that the current through the SCR *will* drop below I_{crit}. To insure that this happens, we require $I_c < I_{crit}$.

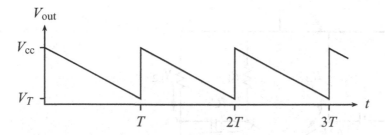

Figure 7.2 The SCR sawtooth oscillator output.

The output voltage for this oscillator is shown in Fig. 7.2. As noted above, the output voltage is initially V_{cc}. Because of the constant charging rate, the capacitor charges linearly and thus V_{out} falls linearly. When the output reaches a voltage low enough to allow the SCR to turn on (labeled V_T), the capacitor discharges quickly (so quickly it looks instantaneous on this scale) and the process starts again.

As with most oscillators, we would like to have an expression for the period of the output waveform. By KVL, the output voltage is $V_{out} = V_{cc} - V_c$. But the capacitor, being charged by the current through the transistor, has voltage $V_c = Q/C = I_c t/C = \beta I_b t/C$. Thus

$$V_{out} = V_{cc} - \beta \frac{I_b t}{C}. \tag{7.3}$$

This equation describes the linear decrease of the signal shown in Fig. 7.2. The period of the signal is defined as the time when $V_{out} = V_T$. Using this in Eq. (7.3) gives

$$T = \frac{C}{\beta I_b} (V_{cc} - V_T) = \frac{R_b C}{\beta (V_{cc} - 0.6)} (V_{cc} - V_T) \tag{7.4}$$

where in the last expression we have used Eq. (7.2) for I_b. It remains to find an expression for V_T. We know that, in order for the SCR to turn on, the output voltage must be less than the gate voltage V_g. How much less will depend on the necessary gate current and R_s. For simplicity, we estimate[3] $V_T \approx V_g$. Using Eq. (7.1) in Eq. (7.4) and simplifying then yields

$$T = \frac{R_b C V_{cc}}{\beta (V_{cc} - 0.6)} \left(\frac{R_1}{R_1 + R_2} \right). \tag{7.5}$$

It is now clear how each circuit parameter affects the waveform period. An easy way to make the period of this oscillator adjustable, for example, would be to make R_b a variable resistor.

---

[3] This is a good approximation since both the gate current and R_s are small.

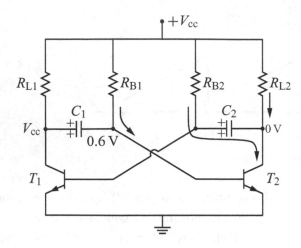

Figure 7.3 The transistor astable oscillator during state 1.

Note how this oscillator has the characteristics of a relaxation oscillator: non-sinusoidal output (in this case, a sawtooth), timing set by capacitor charging, non-linear component operation (in this case, the SCR, which is inherently non-linear, is the relevant component), and analysis in the time domain (we determined V_{out} as a function of time in Eq. (7.3)).

7.2.2 Transistor astable oscillator

Our next relaxation oscillator, the transistor astable oscillator, produces a pulse train output voltage. The two transistors in this circuit, shown in Fig. 7.3, will alternate being saturated ($V_{ce} \approx 0$) or cutoff ($V_{ce} \approx V_{cc}$) with each being in the opposite state of the other. Thus both transistors in this circuit are operating non-linearly. The output voltage can be taken off the collector voltage of either transistor.

We will break the operation of this circuit into four steps involving two stable states and two transitions. Relevant current flows and voltages are shown for the four steps in Figs. 7.3, 7.4, 7.5, and 7.6.

During state 1 of the circuit, transistor T_2 is fully on (i.e., saturated) and T_1 is off (i.e., cutoff). Since T_1 is off, there is no current in resistor R_{L1} and thus no voltage drop across it. The collector voltage of this transistor, V_{c1}, is thus equal to the power supply voltage V_{cc}. Also since T_1 is off, we know its base voltage V_{b1} must be less than the 0.6 V necessary to turn the transistor on. On the other hand, T_2 is on, so its base voltage $V_{b2} \approx 0.6$ V. The base current is supplied through R_{B1}. Since we assume T_2 is saturated, its collector voltage $V_{c2} \approx 0$. The collector current comes from two sources: through R_{L2} and through R_{B2} and C_2. This latter path charges up capacitor C_2.

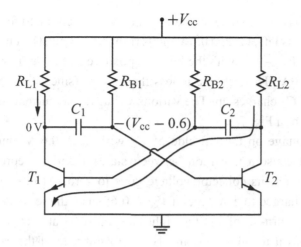

Figure 7.4 The transistor astable oscillator during the transition 1 → 2.

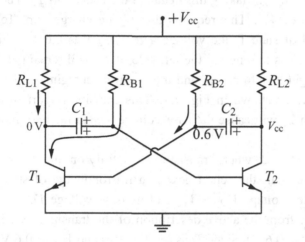

Figure 7.5 The transistor astable oscillator during state 2.

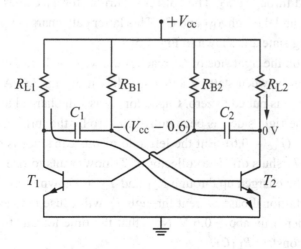

Figure 7.6 The transistor astable oscillator during the transition 2 → 1.

It is important at this point to note the state of the two capacitors. Since capacitor C_1 has voltage V_{cc} on its left side and 0.6 V on its right side, it must be charged with a voltage across it of $V_{cc} - 0.6$, with the left side positive as shown. For capacitor C_2, the right side is 0 V. The left side is less than 0.6 V (since T_1 is off) but is increasing in time as C_2 charges up. The various voltages and currents associated with state 1 are shown in Fig. 7.3.

Eventually, the voltage on the left side of C_2 will reach 0.6 V, thus turning on T_1, and this will cause a transition between states 1 and 2, represented in Fig. 7.4. As T_1 turns on, its collector voltage goes to zero. At this instant, the capacitor C_1 is still charged to a voltage of $V_{cc} - 0.6$; since the left side has been pulled to zero by the turn-on of T_1, the right side must be at $-(V_{cc} - 0.6)$ so that the voltage across it remains constant. But the right side of the capacitor is connected to the base of T_2. Making this negative thus shuts off T_2. The collector of T_2 now wants to be V_{cc}. This requires that C_2 be charged up. To see this, note that at the end of state 1, the voltage across C_2 was 0.6 V, with the left side positive. Since T_1 is turning on, the left side of C_2 will remain at 0.6 V, but now we want the right side to be V_{cc} and this requires charging of the capacitor through R_{L2} and T_1, as shown in Fig. 7.4. This transition spurt of additional base current through T_1 will cause the base voltage V_{b1} to increase momentarily above 0.6 V.

This brings us to state 2, where transistor T_1 is fully on and T_2 is off. Since the circuit is symmetrical, the action now is a mirror image of state 1. Since T_2 is off, its collector voltage $V_{c2} = V_{cc}$ and its base voltage V_{b2} must be less than 0.6 V (actually, from the above description of the transition, we know that $-(V_{cc} - 0.6) < V_{b2} < 0.6$ V). Since T_1 is on, $V_{c1} \approx 0$ and $V_{b1} \approx 0.6$ V, with the base current supplied through R_{B2}. The collector current for T_1 comes from two sources: through R_{L1} and through R_{B1} and C_1. This latter path charges up capacitor C_1. The circuit during state 2 is shown in Fig. 7.5.

When the voltage on the right side of C_1 reaches 0.6 V, T_2 will turn on, and this will cause a transition between states 2 and 1, as shown in Fig. 7.6. As T_2 turns on, its collector voltage is pulled to zero. Capacitor C_2 is still charged to a voltage of $V_{cc} - 0.6$; since the right side has been pulled to zero by the turn-on of T_2, the left side must be at $-(V_{cc} - 0.6)$. But the left side of this capacitor is connected to the base of T_1, so T_1 shuts off. The collector of T_1 now wants to rise to V_{cc} and this requires that C_1 be charged up through R_{L1} and T_2, as shown in Fig. 7.6. This transition spurt of additional base current through T_2 will cause the base voltage V_{b2} to increase momentarily above 0.6 V. Note that the time for this transition is thus set by the time constant $R_{L1}C_1$.

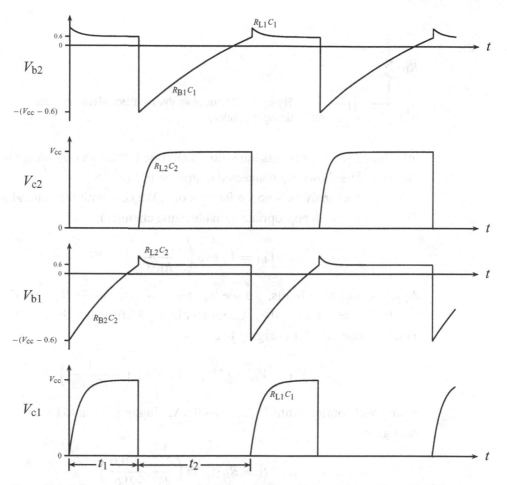

Figure 7.7 Collector and base voltages for the two transistors in the astable oscillator. The RC combinations that determine the time scale for various portions of the curve are noted.

Based on this description, we can sketch the voltage waveforms for V_{c1}, V_{c2}, V_{b1}, and V_{b2} shown in Fig. 7.7. During state 1, V_{c1} is high and V_{c2} is low; V_{b2} is fixed while V_{b1} is increasing with time constant $R_{B2}C_2$. The time constant for the first transition is set by $R_{L2}C_2$. During state 2, V_{c1} is low and V_{c2} is high; V_{b1} is fixed while V_{b2} is increasing with time constant $R_{B1}C_1$. Finally, the time constant for the second transition is set by $R_{L1}C_1$.

We can now exploit our knowledge of the circuit operation and of RC charging to find expressions for the duration of the two states. We have seen that the duration of state 1 is set by the time it takes C_2 to charge (through R_{B2}) from its initial voltage of $-(V_{cc} - 0.6)$ to 0.6 V. The relevant portion of our circuit is shown in Fig. 7.8. The right side of capacitor C_2 is actually connected to the collector of T_2

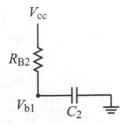

Figure 7.8 Portion of astable oscillator relevant to state 1 timing calculation.

(cf. Fig. 7.3), but since this transistor is saturated its collector voltage is near zero. We have thus shown C_2 connected to ground in Fig. 7.8.

The path of analysis is now a familiar one. We start with the general solution for RC charging (with appropriate variable name changes):

$$V_{b1} = V_1 \exp \left(-\frac{t}{R_{B2}C_2} \right) + V_2. \tag{7.6}$$

Applying our time limits, we see that for $t \to \infty$, $V_{b1} \to V_{cc}$, so $V_2 = V_{cc}$. At $t = 0$, $V_{b1} = -(V_{cc} - 0.6)$. Using this in Eq. (7.6) yields $V_1 = -(2V_{cc} - 0.6)$. Thus our equation for charging becomes

$$V_{b1} = V_{cc} - (2V_{cc} - 0.6) \exp \left(-\frac{t}{R_{B2}C_2} \right). \tag{7.7}$$

State 1 will continue until V_{b1} reaches 0.6 V. Plugging this into Eq. (7.7) and solving for t gives

$$t_1 = R_{B2}C_2 \ln \left(\frac{2V_{cc} - 0.6}{V_{cc} - 0.6} \right). \tag{7.8}$$

A similar analysis gives the duration of state 2:

$$t_2 = R_{B1}C_1 \ln \left(\frac{2V_{cc} - 0.6}{V_{cc} - 0.6} \right) \tag{7.9}$$

so the period of our pulse train is

$$T = t_1 + t_2 = (R_{B1}C_1 + R_{B2}C_2) \ln \left(\frac{2V_{cc} - 0.6}{V_{cc} - 0.6} \right). \tag{7.10}$$

There are two sources of additional restrictions on the circuit components, both of which are based on assumptions we have made along the way. First, we have assumed that the capacitor charging during the transitions will take less time than the duration of the following steady state. This means that

$$t_1 \gg R_{L1}C_1 \quad \text{and} \quad t_2 \gg R_{L2}C_2. \tag{7.11}$$

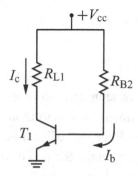

Figure 7.9 Portion of astable oscillator relevant to saturation calculation.

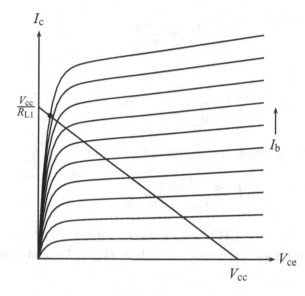

Figure 7.10 Load line plot for saturation calculation. The dot shows the solution when the transistor is saturated. The collector current at this point is I_c^{sat}.

In addition to insuring that our circuit analysis is valid, these inequalities also give output pulses that have less rounding on the leading edge.

The second assumption we have made is that the transistors will be *saturated* when they are on. In order to achieve this, we must be sure the transistor gets enough base current to drive it into saturation. To quantify this, we focus on the portion of our oscillator circuit shown in Fig. 7.9. Our usual KVL analysis then gives $V_{cc} - I_c R_{L1} - V_{ce} = 0$ and $V_{cc} - I_b R_{B2} - V_{be1} = 0$. Solving the first of these for I_c yields

$$I_c = \frac{V_{cc} - V_{ce}}{R_{L1}} \tag{7.12}$$

which is our usual load line equation. This is plotted along with the transistor characteristics in Fig. 7.10.

The second of our voltage loop equations gives

$$I_b = \frac{V_{cc} - V_{be1}}{R_{B2}}.$$ (7.13)

So far our analysis is general. If the transistor is saturated, however, then $I_c = I_c^{sat} \approx V_{cc}/R_{L1}$ (cf. Fig. 7.10). Also, to drive the transistor into saturation, the base current should be more, for a given I_c, than it would be in the linear active region. Thus, $I_b > I_c^{sat}/\beta_1$, where β_1 is the current gain factor for transistor 1. Using Eq. (7.13) for I_b and the above approximation for I_c^{sat}, we obtain

$$\frac{V_{cc} - V_{be1}}{R_{B2}} > \frac{V_{cc}}{\beta_1 R_{L1}}$$ (7.14)

or, rearranging,

$$\beta_1 \left(\frac{V_{cc} - V_{be1}}{V_{cc}} \right) > \frac{R_{B2}}{R_{L1}}.$$ (7.15)

A similar analysis for the circuit driving transistor 2 yields

$$\beta_2 \left(\frac{V_{cc} - V_{be2}}{V_{cc}} \right) > \frac{R_{B1}}{R_{L2}}.$$ (7.16)

Equations (7.11), (7.15), and (7.16), then, place important operating restrictions on our circuit component values in addition to those (Eqs. (7.8) and (7.9)) that set the timing.

7.2.3 The 555 timer

The 555 timer is another "black box" device that can be used for a number of purposes. It comes in an 8-pin package similar to the 741 op-amp. A functional schematic of this device including pin assignments is shown in Fig. 7.11. This is sufficient for us to understand how the device functions. We have three resistors that form a flexible voltage divider, two comparators (basically, op-amps), a transistor, and a new device called a flip-flop. The flip-flop has two outputs, labeled Q and $\bar{Q}$, which have opposite states: when Q is high, $\bar{Q}$ is low, and vice versa. The output Q is available at pin 3, while $\bar{Q}$ is fed to the base of the transistor, thus turning the transistor on or off. The flip-flop also has two inputs, called *set* and *reset*. A high input into *set* causes Q to go high (i.e., to be set), while a low input into *reset* causes Q to go low (i.e., to be reset). The voltage into *set* comes from the output of comparator 1 which compares the voltage on pin 2 with an internal voltage derived

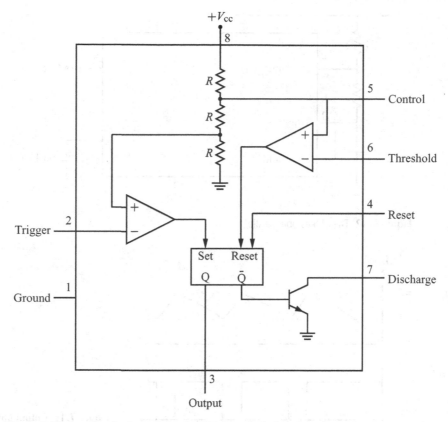

Figure 7.11 Functional schematic for the 555 timer, including pin assignments.

from the voltage divider. The voltage into *reset* can come either from pin 4 or from the output of comparator 2, as shown in the schematic.

A few details: the power supply voltage V_{cc} for the device (input at pin 8) can be anywhere from 4.5 to 16 V. The Output (pin 3) can produce a maximum of 50 mA with a voltage level of $V_{cc} - 2$ V. The Reset (pin 4) is normally held high (defined as above 1.0 V); it must be low (below 0.4 V) to reset the flip-flop. If not used, the Control (pin 5) should be connected to ground via a 0.01 μF capacitor. The Trigger is normally held high (above $\frac{1}{3}V_{cc}$ if the Control is not used) and brought low (below $\frac{1}{3}V_{cc}$ if the Control is not used) to set the flip-flop.

The 555 astable oscillator The 555 is best appreciated by examining some applications. The first of these is an astable oscillator circuit, shown in Fig. 7.12. Its operation is as follows.

1. Upon startup, the capacitor is uncharged so the voltage at pin 2, V_2, is zero. Comparator 1 compares this voltage with $\frac{1}{3}V_{cc}$ coming from the internal voltage

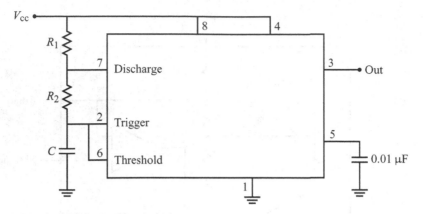

Figure 7.12 The 555 astable oscillator.

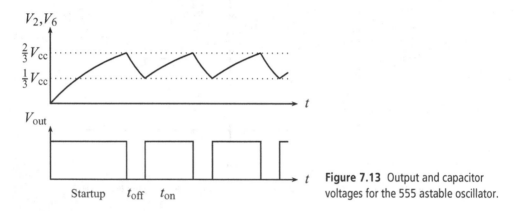

Figure 7.13 Output and capacitor voltages for the 555 astable oscillator.

divider; the resulting high output sets the flip-flop. The output is thus high and the transistor is turned off (since $\bar{Q}$ is low).

2. The capacitor charges with time constant $(R_1 + R_2)C$. When the voltage across the capacitor (pin 6) exceeds $\frac{2}{3}V_{cc}$, comparator 2 outputs a low level which resets the flip-flop. The output then goes low, and the transistor is turned on.

3. The capacitor can now discharge through the transistor with time constant R_2C. The voltage across the capacitor falls until it becomes less than $\frac{1}{3}V_{cc}$, at which point the flip-flop is set again. This charge/discharge cycle then repeats indefinitely.

The output voltage and the voltage across the capacitor for the astable oscillator are shown in Fig. 7.13. Once the circuit has started, the capacitor charges and discharges between $\frac{1}{3}V_{cc}$ and $\frac{2}{3}V_{cc}$, while the output switches between high (during charge) and low (during discharge). It can be shown that the duration of the high

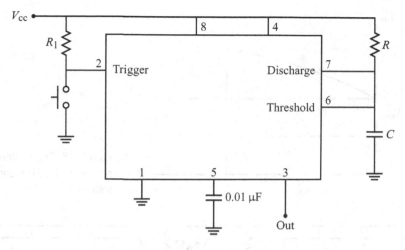

Figure 7.14 The 555 used as a timer.

and low portions of the output waveforms are given by

$$t_{\text{on}} = (R_1 + R_2)C \ln 2 \quad \text{and} \quad t_{\text{off}} = R_2 C \ln 2. \tag{7.17}$$

555 monostable operation (timer mode) As a second example, we consider the use of the 555 as a timer.[4] The circuit is shown in Fig. 7.14 and operates as follows.

1. Normally, the flip-flop is reset. This means that the output is low, and the transistor is turned on, so the capacitor is uncharged.
2. When the button is pushed, the Trigger input is grounded. This sets the flip-flop, makes the output high, and turns off the transistor.
3. Now the capacitor can charge up. It charges with time constant RC until $V_6 > \frac{2}{3}V_{\text{cc}}$, at which point the flip-flop is reset and C is quickly discharged. The circuit is thus left in its starting state.

The relevant waveforms for this circuit are shown in Fig. 7.15. A single pulse is produced at the output of the 555. This can be used to set the duration of some other process. The length of this pulse is set by the RC charging of the capacitor and is given by

$$t_{\text{on}} = RC \ln 3. \tag{7.18}$$

---

[4] Strictly, this circuit is out of place here since it is not a relaxation oscillator, but discussing it here will give some flavor of the versatility of the 555.

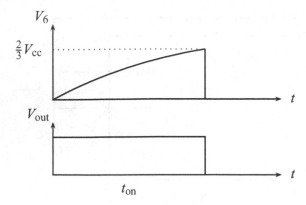

Figure 7.15 Output and capacitor voltages for the 555 monostable oscillator.

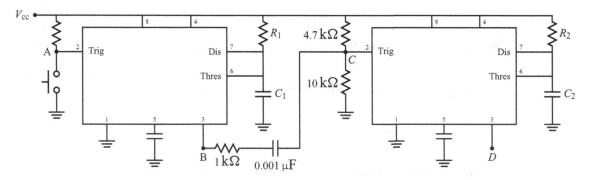

Figure 7.16 Cascading two 555 timers.

For some applications (e.g., producing a sequence of timed events) we would like to start a second timer at the end of the pulse output by a first timer. Figure 7.16 shows a method to cascade two or more timers. The first timer (on the left) is configured as in Fig. 7.14. The configuration of the second timer is similar, but its Trigger input is held at $0.68V_{cc}$ by the voltage divider formed by the 4.7 kΩ and 10 kΩ resistors. Note that this value is set well above the level necessary to trigger the second timer (i.e., $\frac{1}{3}V_{cc}$). The output of timer 1 is also connected to the Trigger of timer 2 through the 1 kΩ resistor and the 0.001 μF capacitor. The capacitor, along with the circuit resistors, forms a differentiator. The negative spike resulting from the differentiation of the timer 1 output is enough to pull the Trigger input of timer 2 below $\frac{1}{3}V_{cc}$, and thus timer 2 is activated. The relevant waveforms for this process are shown in Fig. 7.17.

Two final notes on the use of the 555. It is sometimes desirable to invert the output voltage of the 555. A simple transistor inverter (see, e.g., Section 4.3) will do the job. The output of the 555 is connected to the transistor base through a resistor and the inverted signal is taken from the collector of the transistor. The

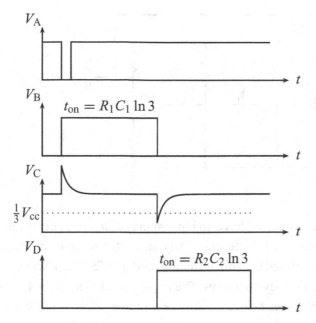

Figure 7.17 Waveforms for the cascading circuit.

base resistor must be chosen small enough to insure that the transistor is saturated by the 555 output.

None of our applications has used the Control input of the 555. Referring to Fig. 7.11, we see that placing a voltage V_{con} on the Control input changes the comparison voltage of comparator 1 (i.e., the trigger level) to $\frac{1}{2}V_{con}$ and the comparison voltage of comparator 2 (i.e., the threshold level) to V_{con}. We will leave it as an exercise to explore how these changes affect the astable oscillator operation.

7.3 Sinusoidal oscillators

We now turn to the other major class of oscillators, the *sinusoidal oscillators*. Since these oscillators all have a sinusoidal output, our analysis focuses on determining the frequency of this sine wave. It is useful for sinusoidal oscillators to employ some of the ideas we developed when discussing feedback in Section 4.4.9. We saw that the gain of an amplifier with feedback was

$$a' = \frac{a}{1 - a\beta} \tag{7.19}$$

where a was the gain of the amplifier without feedback, and β was the feedback ratio or the fraction of the amplifier output that was added to the input.

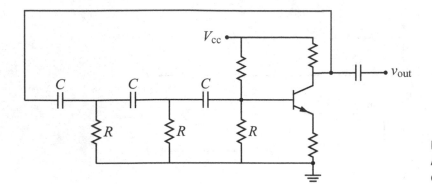

Figure 7.18
A simple RC oscillator.

For an oscillator circuit, we have no input signal. If $a\beta = 1$, however, Eq. (7.19) says that our circuit has infinite gain. Any small fluctuation on the output will automatically be amplified by a huge amount, and this feedback will continue until the circuit reaches a steady condition. The requirement that $a\beta = 1$ for oscillation is called the *Barkhausen criterion*. As we will see, β is usually a function of frequency and the Barkhausen criterion is thus only satisfied for one frequency. That frequency, then, will be the frequency of our output waveform.

7.3.1 RC oscillator

Our first example of a sinusoidal oscillator is the *RC oscillator* shown in Fig. 7.18. The amplifier portion of this circuit should look familiar – it is the common-emitter transistor amplifier. Recall that the voltage gain of this amplifier is given by

$$a = \frac{v_{out}}{v_{in}} = \frac{-\beta(R_c \| R_L)}{r_{be} + (\beta + 1)R_e}. \tag{7.20}$$

The minus sign in this expression tells us that the output of this amplifier is inverted or, for a sinusoidal signal, shifted by 180 degrees. Since we want $a\beta$ to be positive to satisfy the Barkhausen criterion, our feedback circuit must introduce another 180 degree phase shift.

We employ a three-stage RC high-pass filter for our feedback network. Recall from our early studies of RC circuits that a high-pass filter also acts as a positive phase shifter, with the phase shift given by $\phi = \tan^{-1}(1/\omega RC)$. For the high-pass filter, this phase is between zero and 90 degrees, so it is reasonable to expect that three of these could produce 180 degrees of phase shift if the frequency was right.

For a more rigorous analysis, consider the feedback network alone, as shown in Fig. 7.19. We use the mesh loop method, employing the current loops indicated.

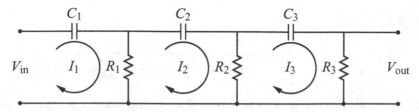

Figure 7.19 The feedback network for the RC oscillator.

These produce the following three equations:

$$V_{in} = I_1 \left(\frac{1}{j\omega C_1} \right) + (I_1 - I_2)R_1 \tag{7.21}$$

$$0 = I_2 \left(\frac{1}{j\omega C_2} \right) + (I_2 - I_3)R_2 + (I_2 - I_1)R_1 \tag{7.22}$$

$$0 = I_3 \left(\frac{1}{j\omega C_3} \right) + I_3 R_3 + (I_3 - I_2)R_2. \tag{7.23}$$

For the oscillator of Fig. 7.18, all the resistor values and all the capacitor values are the same. Dropping the subscripts and rearranging, we obtain

$$\left(R + \frac{1}{j\omega C} \right) I_1 - RI_2 = V_{in} \tag{7.24}$$

$$-RI_1 + \left(2R + \frac{1}{j\omega C} \right) I_2 - RI_3 = 0 \tag{7.25}$$

$$-RI_2 + \left(2R + \frac{1}{j\omega C} \right) I_3 = 0. \tag{7.26}$$

These latter three equations have been arranged to emphasize the fact that they form a set of linear, simultaneous equations for the three unknowns I_1, I_2, and I_3.

We are interested in finding β, the feedback ratio for this network. Now since β is the ratio of the output voltage to the input voltage, and since the output voltage is $I_3 R$, we need only solve our equations for I_3. Brute force algebra or Cramer's Method of Determinants (see Appendix B) can be employed to yield

$$I_3 = \frac{V_{in}R^2}{\left(R + \frac{1}{j\omega C} \right) \left[\left(2R + \frac{1}{j\omega C} \right)^2 - R^2 \right] - R^2 \left(2R + \frac{1}{j\omega C} \right)}$$

$$= \frac{V_{in}R^2}{R^3 - 5\frac{R}{\omega^2 C^2} + \frac{1}{j\omega C} \left[6R^2 - \frac{1}{\omega^2 C^2} \right]} \tag{7.27}$$

and

$$\beta = \frac{V_{out}}{V_{in}} = \frac{I_3 R}{V_{in}} = \frac{1}{1 - \frac{5}{(\omega RC)^2} + j\left[\frac{1}{(\omega RC)^3} - \frac{6}{\omega RC}\right]}. \tag{7.28}$$

Now, recall that our goal is to make $a\beta = 1$, and since a is negative and real for the common-emitter amplifier, β must also be negative and real. To be real, the imaginary part must be zero:

$$\frac{1}{(\omega RC)^3} - \frac{6}{\omega RC} = 0 \tag{7.29}$$

which is satisfied only if

$$\omega = \frac{1}{\sqrt{6}RC} \equiv \omega_0 \tag{7.30}$$

so the oscillator will have this frequency. Plugging this ω back into Eq. (7.28), we get $\beta = -\frac{1}{29}$. Thus, to satisfy the Barkhausen criterion, our amplifier must be designed to have a voltage gain $a = -29$.

7.3.2 Oscillator stability

An important issue when evaluating the merit of an oscillator is *frequency stability*. We have shown for the RC oscillator that the Barkhausen criterion is satisfied exactly for one frequency ω_0. In practice, the frequency of the oscillator will tend to drift away from ω_0. Eventually, the circuit senses something is wrong (because the phase shift is no longer 180 degrees) and returns the oscillator frequency back to ω_0.

For most applications, this frequency drift is undesirable, so when evaluating an oscillator we would like to know how much the oscillator can drift before it corrects itself. It is useful in this regard to plot the phase shift of our feedback network as a function of frequency. From Eq. (7.28), the phase ϕ of the output signal relative to the input is given by

$$\tan\phi = \frac{\text{Im}(\beta)}{\text{Re}(\beta)} = \frac{\frac{1}{(\omega RC)^3} - \frac{6}{\omega RC}}{1 - \frac{5}{(\omega RC)^2}} = \frac{1 - \left(\frac{\omega}{\omega_0}\right)^2}{\frac{1}{\sqrt{6}}\left(\frac{\omega}{\omega_0}\right)\left[\frac{1}{6}\left(\frac{\omega}{\omega_0}\right)^2 - 5\right]}. \tag{7.31}$$

A plot of the relative phase (i.e., $\phi - 180$) versus the scaled frequency ω/ω_0 is shown by the curve labeled "Simple RC" in Fig. 7.20. The relatively weak dependence

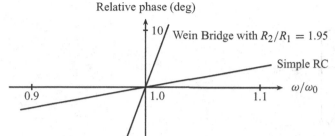

Figure 7.20 Stability plot for the simple RC and Wein bridge networks.

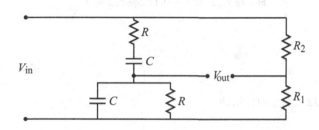

Figure 7.21 The RC Wein bridge feedback network.

of ϕ on ω means that the oscillator frequency can drift quite a bit before the phase changes enough for the oscillator to correct itself, so the stability of this feedback network is not very good.

7.3.3 RC Wein bridge oscillator

A feedback network with better stability characteristics is the *RC Wein bridge*, shown in Fig. 7.21. An analysis of this network gives

$$\beta = \frac{1}{3 + j\left(\frac{\omega}{\omega_0} - \frac{\omega_0}{\omega}\right)} - \frac{1}{1 + \frac{R_2}{R_1}} \tag{7.32}$$

where $\omega_0 = 1/RC$. The phase difference ϕ is given by

$$\tan\phi = \frac{-\left(1 + \frac{R_2}{R_1}\right)\left(\frac{\omega}{\omega_0} - \frac{\omega_0}{\omega}\right)}{3\frac{R_2}{R_1} - \left(\frac{\omega}{\omega_0} - \frac{\omega_0}{\omega}\right)^2 - 6}. \tag{7.33}$$

The relative phase given by Eq. (7.33) is also plotted in Fig. 7.20 where, for example, we have chosen the case $R_2/R_1 = 1.95$. Note that here the phase varies much more rapidly with ω than for the simple RC network. This means that an oscillator using this network will have better frequency stability than one using the network of Fig. 7.19.

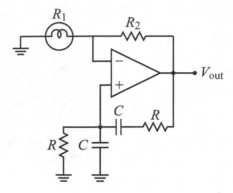

Figure 7.22 The Wein bridge oscillator.

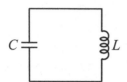

Figure 7.23 A simple LC circuit.

An example of such usage is shown in Fig. 7.22. In this case the circuit amplifi-cation is provided by an op-amp. The circuit also employs an interesting technique to control the output *amplitude*. R_1 is a light bulb, and R_2 is selected so that $R_2 = (2 + \epsilon)R_1$, where ϵ is a small number. Under these conditions, it can be shown that Eq. (7.32) reduces to $\beta \approx \frac{1}{9}\epsilon$. If the output voltage increases, the current through the light bulb increases and heats the bulb filament. This increases R_1 and, since R_2 is fixed, ϵ must decrease. The feedback ratio β is thus decreased, which in turn *decreases* the output, counteracting the original increase. Similarly, if the output decreases, cooling of the filament will increase ϵ and thus increase the output. This technique thus stabilizes the output amplitude.

7.3.4 LC tank circuit oscillators

Another type of sinusoidal oscillator is the *LC* or *tank circuit oscillator*. This type of oscillator is based on the fact that a simple LC circuit (also called a tank circuit), such as that shown in Fig. 7.23, will sustain oscillations. Applying KVL to this circuit produces

$$\frac{Q}{C} + L\frac{dI}{dt} = 0. \tag{7.34}$$

Taking the derivative then gives

$$\frac{d^2I}{dt^2} + \frac{1}{LC}I = 0. \tag{7.35}$$

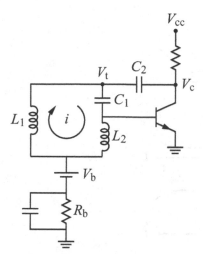

Figure 7.24 A Hartley oscillator.

This is the harmonic oscillator equation with well-known solution $I = I_0 \sin(\omega_0 t + \phi)$, where $\omega_0 = 1/\sqrt{LC}$. Without circuit resistance, the current will oscillate forever. Of course, all real circuits have at least a small resistance. In this case, the oscillation frequency will still be approximately ω_0, but the oscillations will die away in time. To use this circuit to make a practical oscillator, we need to figure out how to "kick" the circuit periodically in such a way that the oscillations are maintained.

7.3.4.1 The Hartley oscillator

One example of such a circuit is the Hartley oscillator shown in Fig. 7.24. The DC voltage V_b and the resistance R_b produce a constant base current so the transistor remains in the linear active region. As usual, we treat the DC voltage and the capacitor spanning R_b as AC shorts, so the bottom of the tank circuit formed by L_1, L_2, and C_1 is at AC ground.

Suppose that an oscillating current i is flowing in the tank circuit as shown. Then $v_{be} = j\omega L_2 i$ (here we continue the convention of using lower case for AC quantities while upper case will indicate the total signal). As the current oscillates, so does V_{be}, as shown in Fig. 7.25. The base current also oscillates, but due to the non-linear relationship between I_b and V_{be}, the current is not sinusoidal, but exhibits a spike when V_{be} is maximum. This in turn causes a spike in the collector current and a downward spike in the collector voltage V_c. This couples through C_2 and causes a depression in the voltage V_t at the top of the tank circuit. This happens at just the right time to reinforce the oscillations, just like pushing a child on a swing. To see this last point in another way, note that $v_t = -j\omega L_1 i = -(L_1/L_2)v_{be}$.

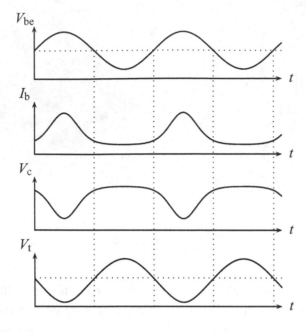

Figure 7.25 Selected voltages for the Hartley oscillator.

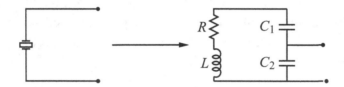

Figure 7.26 Circuit symbol for the piezoelectric crystal, and its equivalent circuit.

The minus sign means that v_t is 180 degrees out of phase with v_{be}, as shown in Fig. 7.25, so the reinforcing depression in v_t does occur at the needed time (i.e., when v_t is at its minimum).

7.3.4.2 Crystal oscillators

A piezoelectric crystal is often used in LC oscillators. The circuit symbol for this device is shown in Fig. 7.26 along with its equivalent circuit. Note that the equivalent circuit is essentially an LC tank circuit. An example of an oscillator using a piezoelectric crystal is shown in Fig. 7.27.

Different oscillator circuits are used for different purposes. Some of the characteristics of different feedback networks are shown in Table 7.1. Crystal oscillators are commonly used in a wide variety of applications due to their excellent stability and low cost, but they are not suitable for high power applications. The RC and LC feedback networks are used for such cases.

Table 7.1 Comparison of feedback network characteristics

Feedback circuit	Typical frequency range	Stability $(\delta\omega/\omega)$
RC Wein bridge	5 Hz – 1 MHz	10^{-3}
LC tank circuit	10 kHz – 100 MHz	10^{-4}
Crystal	10 kHz – 100 MHz	$10^{-6} - 10^{-8}$

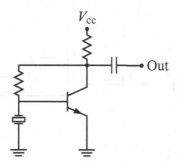

Figure 7.27 The Pierce crystal oscillator.

7.4 Oscillator application: EM communications

7.4.1 Introduction

A common way to communicate information is through electromagnetic (EM) waves. Examples include radio and television waves, telephone microwave links, light waves in optical fibers, radar, and cell phone signals. There are various ways of sending information on an EM wave. For analog signals, the most common techniques are *amplitude modulation* or *AM* and *frequency modulation* or *FM*. These techniques are also used when studying wave phenomena in the research lab.

7.4.2 Amplitude modulation

As the name implies, for amplitude modulation we take a constant frequency sine wave (called the *carrier wave*) and vary the amplitude with time. The "information" being communicated is in the amplitude variation. The amplitude variation is usually at a much lower frequency than the carrier wave. For example, with AM radio waves, the carrier frequency is in the range 0.53–1.60 MHz, while the audio information that modulates the carrier is in the range 20–20 000 Hz. A typical AM wave is shown in Fig. 7.28.

In electronics, two questions are of interest: (1) how does one create an AM sine wave and (2) how does one extract the information from (or *demodulate*) an AM

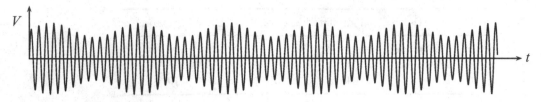

Figure 7.28 An amplitude modulated sine wave.

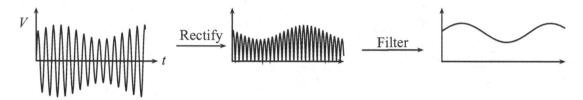

Figure 7.29 Demodulating a low frequency AM wave.

sine wave? The answers depend on the frequency of the carrier wave. Here we will give examples for relatively low frequency signals (e.g., radio waves) and for high frequency signals (light waves).

7.4.2.1 Low frequency AM

The process of demodulating a low frequency AM wave is fairly straightforward and is shown schematically in Fig. 7.29. The received signal is first rectified so that the modulated signal is always positive. Rectification is followed by a low-pass filter that smooths the signal out, thus leaving the modulation. The process is not unlike that used when creating a power supply, but here special "fast response" diodes are used. The modulation signal can then be capacitively coupled to an amplifier stage and the information it carries extracted.

The process of creating the AM signal is more complicated. A signal like that shown in Fig. 7.28 is produced by the function

$$V = V_1(1 + m \sin \omega_m t) \sin \omega_c t \tag{7.36}$$

where ω_c is the carrier frequency, ω_m is the modulation frequency, and $m \leq 1$ is the modulation index. Using some trig identities, Eq. (7.36) can be written as

$$V = V_1 \left[\sin \omega_c t + \frac{m}{2} \cos (\omega_c - \omega_m)t - \frac{m}{2} \cos (\omega_c + \omega_m)t \right]. \tag{7.37}$$

From this form we can see that our AM sine wave consists of cosine waves at *three* different frequencies: ω_c, $\omega_c + \omega_m$, and $\omega_c - \omega_m$. We could thus, in principle,

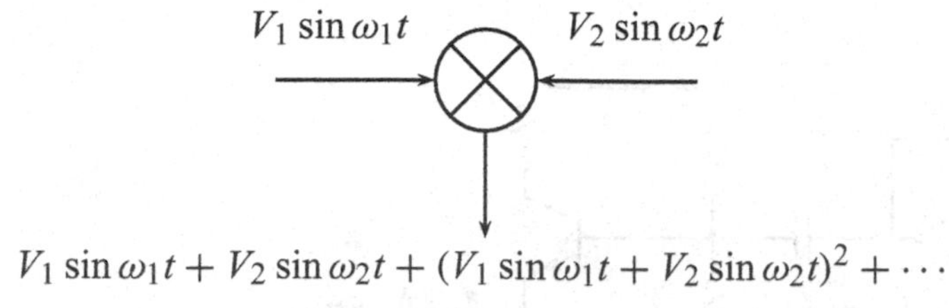

$$V_1 \sin \omega_1 t \quad\quad V_2 \sin \omega_2 t$$

$$V_1 \sin \omega_1 t + V_2 \sin \omega_2 t + (V_1 \sin \omega_1 t + V_2 \sin \omega_2 t)^2 + \cdots$$

Figure 7.30 The symbol and functioning of a mixer.

produce our AM sine wave by adding together three signals at these frequencies. In most cases, however, a device known as a *mixer* is used. A mixer is a device with two inputs which gives an output which is proportional to the sum of the original two signals plus the square of the sum of the two signals.[5] This is shown schematically in Fig. 7.30.

Again using a series of trig identities, we can rewrite the squared term in the mixer output as follows:

$$
\begin{aligned}
f &= (V_1 \sin \omega_1 t + V_2 \sin \omega_2 t)^2 \\
&= V_1^2 \sin^2 \omega_1 t + V_2^2 \sin^2 \omega_2 t + 2V_1 V_2 \sin \omega_1 t \sin \omega_2 t \\
&= V_1^2 (1 - \cos^2 \omega_1 t) + V_2^2 (1 - \cos^2 \omega_2 t) \\
&\quad + V_1 V_2 \left[\cos (\omega_1 - \omega_2)t - \cos (\omega_1 + \omega_2)t \right] \\
&= V_1^2 \left[1 - \frac{1}{2}(1 + \cos 2\omega_1 t) \right] + V_2^2 \left[1 - \frac{1}{2}(1 + \cos 2\omega_2 t) \right] \\
&\quad + V_1 V_2 \left[\cos (\omega_1 - \omega_2)t - \cos (\omega_1 + \omega_2)t \right]
\end{aligned}
\tag{7.38}
$$

all of which is added to the original signals. The point of this complicated mess is that this last expression includes terms that involve the sum and difference of our two input frequencies. If we filter out the frequencies we do not want (ω_2, $2\omega_1$, and $2\omega_2$), we are left with

$$V = V_1 \sin \omega_1 t + V_1 V_2 \left[\cos (\omega_1 - \omega_2)t - \cos (\omega_1 + \omega_2)t \right] \tag{7.39}$$

which is the same form as Eq. (7.37). Thus a mixer allows us to take two sine waves and produce an amplitude modulated wave.

7.4.2.2 High frequency AM

One can also transmit AM signals via light waves and detect these waves using solid-state detectors. An example of such a system is shown in Figs. 7.31 and

---

[5] Note that this is not the same device as the audio mixers often used in musical productions.

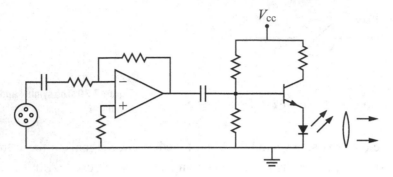

Figure 7.31 AM lightwave transmitter.

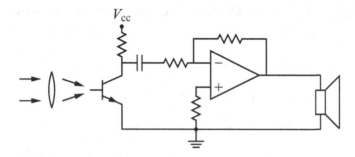

Figure 7.32 AM lightwave receiver.

7.32. The carrier wave in this case is the red light from the LED and the amplitude modulation is the variation of the light intensity. The transmitter uses a transistor circuit to maintain DC current through an LED. This current (and, thus, the LED light intensity) is then modulated by adding an AC component to the base current. In this case, the AC component is simply the audio output of a microphone, amplified by the op-amp portion of the circuit.

The receiver portion of this system (see Fig. 7.32) uses a photo-transistor. The photo-transistor is an example of a light sensitive device; other such devices include the photo-diode and the solar cell. These devices work by absorbing photons which then promote electrons into the conduction band. If this promotion takes place in the base portion of a transistor, these electrons can take the place of emitter electrons that have recombined with holes, thus increasing the collector current. In our circuit we focus the transmitted light on the photo-transistor and then amplify the resulting collector modulations and directly receive the transmitted audio signal. There is no need for a mixer or rectification/filtering in this system.

Figure 7.33 A frequency modulated sine wave.

7.4.3 Frequency modulation

A second way to transmit information in an EM wave is through frequency modulation (FM). In this technique, the amplitude of the wave is kept constant, but the frequency of the wave varies with time. A typical FM wave is shown in Fig. 7.33. Such frequency modulation can be represented by an equation of the form

$$V = V_0 \cos \left[\omega_1 - \omega_2(t) \right] t \tag{7.40}$$

where ω_1 is the carrier wave frequency and ω_2 is a time-varying frequency that contains the information to be transmitted (e.g., the audio of an FM radio wave).

A mixer is used both to produce and to demodulate the FM signal. We have seen in Eq. (7.38) that if we mix two signal of frequencies ω_1 and ω_2, we obtain several output frequencies, including $\omega_1 - \omega_2$. We thus select this frequency using a band-pass filter that is narrow enough to reject all the other frequencies. This frequency is then transmitted and received at a remote location. At this location, the received signal at frequency $\omega_1 - \omega_2$ is mixed with another signal of frequency ω_1. Again, the mixer output has several frequencies, including one that is the difference between the two input frequencies: $\omega_1 - (\omega_1 - \omega_2)$ or ω_2. We again filter out all the frequencies except ω_2 and our original information signal is recovered.

EXERCISES

1. Explain the operation of the circuit in Fig. 7.34. Include a description of its output waveform, including its amplitude and period.

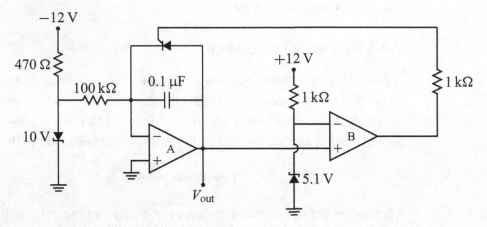

Figure 7.34 Circuit for Problem 1.

2. Derive the expression $t_{on} = (R_1 + R_2)C \ln 2$ for the 555 astable oscillator.
3. Design a system, using the 555 timer, that will turn your lights on for four hours each day while you are away on vacation. Assume that a high logic level will be used to turn the lights on.
4. Design a timing circuit for a home security system that will activate the system 30 seconds after you push a button (giving you time to leave the house) and then de-activate 8.5 hours later (right before you arrive home). Assume that a low logic level activates the system.
5. Consider the relaxation oscillator of Fig. 7.35. Derive an expression for the period of this oscillator in terms of R, C, V_{in}, V_{sat}^+, and V_{sat}^-. Hint: compare this

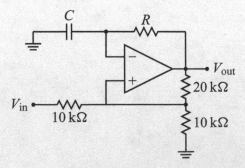

Figure 7.35 Circuit for Problem 5.

circuit with that of Section 6.5; start your analysis by finding an expression for the voltage at the (+) input of the op-amp in terms of V_{in} and V_{out}.

6. Suppose you take the 555 astable oscillator circuit and place a constant voltage V_{con} on the Control input. Find expressions for t_{on} and t_{off} for this new circuit.

7. Consider the feedback network shown in Fig. 7.36.

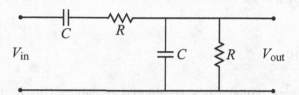

Figure 7.36 Circuit for Problem 7.

(a) At what frequency ω will the network give a real value for the feedback ratio β?

(b) Which transistor amplifier configuration could be used with this network to make an oscillator circuit?

(c) What voltage gain a would be required for this circuit?

FURTHER READING

Irving M. Gottlieb, *Understanding Oscillators* (Indianapolis, IN: Sams, 1971).

Richard J. Higgins, *Electronics with Digital and Analog Integrated Circuits* (Englewood Cliffs, NJ: Prentice-Hall, 1983).

Paul Horowitz and Winfield Hill, *The Art of Electronics*, 2nd edition (New York: Cambridge University Press, 1989).

Walter G. Jung, *IC Timer Cookbook* (Indianapolis, IN: Sams, 1978).

8 Digital circuits and devices

8.1 Introduction

In analog electronics, voltage is a continuous variable. This is useful because most physical quantities we encounter are continuous: sound levels, light intensity, temperature, pressure, etc.[1] Digital electronics, in contrast, is characterized by only two distinguishable voltages. These two states are called by various names: on/off, true/false, high/low, and 1/0. In practice, these two states are defined by the circuit voltage being above or below a certain value. For example, in TTL logic circuits, a high state corresponds to a voltage above 2.0 V, while a low state is defined as a voltage below 0.8 V.[2]

The virtue of this system is illustrated in Fig. 8.1. We plot the voltage level versus time for some electronic signal. If this was part of an analog circuit, we would say that the voltage was averaging about 3 V, but that it had, roughly, a 20% noise level, rather large for most applications and thus unacceptable. For a TTL digital circuit, however, this signal is always above 2.0 V and is thus always in the high state. There is no uncertainty about the digital state of this voltage, so the digital signal has zero noise. This is the primary advantage of digital electronics: it is relatively immune to the noise that is ubiquitous in electronic circuits. Of course, if the fluctuations in Fig. 8.1 became so large that the voltage dipped below 2.0 V, then even a digital circuit would have problems.

8.2 Binary numbers

Although digital circuits have excellent noise immunity, they also are limited to producing only two levels. This does not appear to be very helpful in representing the continuous signals we so frequently encounter. The solution starts with the

---

[1] This holds for most macroscopic quantities. On the atomic level, many physical quantities are quantized.
[2] If the voltage is between these thresholds, we say the state is undetermined, which means the circuit behavior cannot be insured.

Table 8.1 The first twelve counting numbers in binary

Base 10	Base 2	Base 10	Base 2
0	0	6	110
1	1	7	111
2	10	8	1000
3	11	9	1001
4	100	10	1010
5	101	11	1011

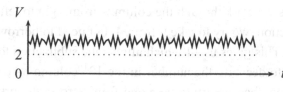

Figure 8.1 A noisy analog signal is noise-free in digital.

realization that we can represent a signal level by a number that only uses two digits. For these *binary numbers*, the two digits used are 0 and 1. Binary numbers are also call *base 2* numbers, and can be understood by abstracting the rules we all know for the numbers we commonly use (base 10 numbers). When we write down a base 10 number, each digit can have 10 possible values, 0 to 9, and each digit corresponds to 10 raised to a power. For example, when I write 1024_{10}, this is equal to

$$1024_{10} = 1 \times 10^3 + 0 \times 10^2 + 2 \times 10^1 + 4 \times 10^0 \tag{8.1}$$

where we use the subscript 10 on 1024 to make explicit the base of the number.

Analogously, for binary numbers, each digit can have only 2 possible values, 0 or 1, and each digit of the number corresponds to 2 raised to a power. Thus

$$10110_2 = 1 \times 2^4 + 0 \times 2^3 + 1 \times 2^2 + 1 \times 2^1 + 0 \times 2^0 = 22_{10}. \tag{8.2}$$

The first twelve base 10 numbers and their binary equivalents are given in Table 8.1.

In a similar manner, the rules for base 10 addition and subtraction can be mapped over to binary arithmetic. Some examples are shown in Table 8.2. In base 10, when we add the rightmost column, 9 plus 5 equals 14. Since this result cannot be expressed in a single digit with the ten available digits (0 to 9), we write down the 4 and carry the 1 to the next column. Similarly, when we add the 1 and 1 of the rightmost column of the binary number, we get 10_2. Since this cannot be expressed

Table 8.2 Adding and subtracting in binary

Addition		Subtraction	
Base 10	Base 2	Base 10	Base 2
9	1001	14	1110
+15	+1111	−9	−1001
24	11000	5	101

in a single binary digit, we write down the 0 and carry the 1 to the next column. We repeat the process as we work through the columns from right to left.

In the base 10 subtraction, we try to take 9 from 4, but need to borrow from the next column to the left. This gives us 14 which allows the subtraction to proceed, but we must remember to decrease the number in the 10^1 column by one. In like manner, for the binary number, we try to take one from zero in the first column. To progress, we need to borrow from the next column (the 2^1 column), carefully decreasing that column's digit by one. We then continue as with the base 10 number until we reach the leftmost column.

Note from Eq. (8.2) and Tables 8.1 and 8.2 that binary numbers tend to be long (i.e., have many digits) compared to base 10 numbers. As we will see, this has a direct impact on the complexity of digital circuits.

8.3 Representing binary numbers in a circuit

In the last section we saw that numbers can be expressed in base 2 just as they can in the more familiar base 10. Base 2 is particularly suitable for expressing a number digitally since digital electronics has only two levels, high and low, and these can be taken to represent the two digits (1 and 0) in a binary number. But a binary number typically consists of several digits.[3] How can we express all of these digits electronically? There are two basic methods, known as *parallel* and *serial* representation.

In parallel expression of a binary number, each digit or bit of the number is represented *simultaneously* by a voltage in the circuit. This is represented schematically in Fig. 8.2. Each output line has a high or low voltage relative to ground. These lines are assigned to represent a particular bit of the number. In this example,

---

[3] A *binary digit* is often referred to as a *bit*.

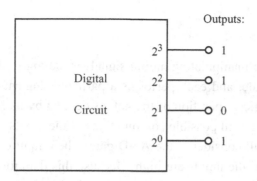

Figure 8.2 Parallel representation of a four-bit number 1101.

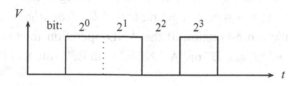

Figure 8.3 Serial representation of a four-bit number 1011.

the bottom line represents the 2^0 bit, the next line up represents the 2^1 bit, and so on. Because we have four independent lines, the entire four-bit number can be expressed at a point in time, so parallel communication of information is very fast. The price we pay for this speed is the increased number of lines in our circuit. The more precision we want in our number, the more significant figures we need, and the more lines are required.

An alternative way of expressing a binary number is by a serial representation. In this method, the various bits are communicated by sending a *time sequence* of high/low voltage levels on a single line. An example of this is shown in Fig. 8.3. The plot shows the voltage level on a serial line. The voltage switches between high and low levels, with each level lasting for a certain time interval. The first interval corresponds to the 2^0 bit of our number, the next interval represents the 2^1 bit, and so on. We are thus able to communicate the binary number on a single line, rather than the multiple lines required for parallel communications, but the communication is no longer instantaneous; we must wait for several intervals before we receive all the bits of our transmitted number.

In order for serial transmission of information to work, both the sender and receiver need to agree about several things. Some of these are: (1) how many bits of data are going to be sent, (2) what digital level (high or low) corresponds to the 1 bit, (3) what is the time interval between bits, and (4) how will the start of a number be recognized?

8.4 Logic gates

The basic circuit element for manipulating digital signals is the *logic gate*. There are several types of logic gate, and each performs a particular logical operation on the input signals. The logical operation of the gate is defined by its *truth table* which gives the output state for all possible combinations of the inputs.

The first logic gate we will consider is the AND gate. The output of an AND gate is high only when all of the inputs are high. Because this definition is clear for any number of inputs, this type of gate can, in principle, have as many inputs as you like. In Fig. 8.4 we show a two-input AND gate along with its truth table. As we will see below, there is a special algebra, called *Boolean algebra*, for logic operations. The symbolic representation of the AND operation for two inputs A and B is $A \cdot B$, pronounced "A and B" or "A ANDed with B." Note that it is *not* "A times B."

The output of an OR gate is high when any input is high. Again, this operation is defined for any number of inputs. A two-input OR gate is shown in Fig. 8.5 along with its truth table and logical expression $A + B$. $A + B$ is pronounced "A or B" or "A ORed with B." It is *not* "A plus B."

A third gate is called the exclusive-OR gate, or simply the XOR gate. The logic here is that the output will be high when either input is high, but not when both inputs are high. Note that this definition assumes there are only two inputs. This device is shown in Fig. 8.6. The circuit symbol is like the OR gate symbol; the curved line across the inputs denotes the exclusion described in the definition. The circle around the + in the Boolean expression $A \oplus B$ distinguishes this expression from the OR function. This is pronounced "A x or B."

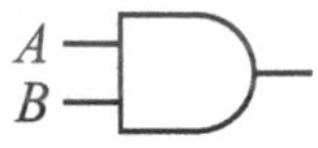

A	B	Out
0	0	0
1	0	0
0	1	0
1	1	1

A
B — Out $\equiv A \cdot B$

Figure 8.4 Circuit symbol, algebraic expression, and truth table for the AND gate.

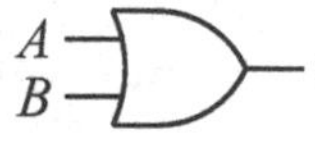

A	B	Out
0	0	0
1	0	1
0	1	1
1	1	1

A
B — Out $\equiv A + B$

Figure 8.5 Circuit symbol, algebraic expression, and truth table for the OR gate.

A ⎤
B ⎦ ⟩—Out ≡ $A \oplus B$

A	B	Out
0	0	0
1	0	1
0	1	1
1	1	0

Figure 8.6 Circuit symbol, algebraic expression, and truth table for the XOR gate.

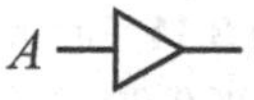

A —▷— Out ≡ A

A	Out
0	0
1	1

Figure 8.7 Circuit symbol, algebraic expression, and truth table for the buffer gate.

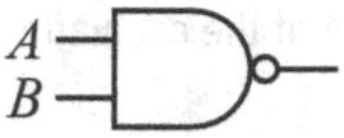

A ⎤
B ⎦ D∘— Out ≡ $\overline{A \cdot B}$

A	B	Out
0	0	1
1	0	1
0	1	1
1	1	0

Figure 8.8 Circuit symbol, algebraic expression, and truth table for the NAND gate.

A ⎤
B ⎦ ⟩∘— Out ≡ $\overline{A + B}$

A	B	Out
0	0	1
1	0	0
0	1	0
1	1	0

Figure 8.9 Circuit symbol, algebraic expression, and truth table for the NOR gate.

The buffer gate, shown in Fig. 8.7, seems to be superfluous. It has only one input, and the output is the same as the input. What good is this? This gate is used to regenerate logic signals. A logical high signal may start out at 5 V, say, but after being transmitted on conductors with non-zero resistance or after driving several other logic gates, the voltage level may fall and become perilously close to the defining threshold for a high logic level. The buffer is then used to boost the level up to a healthier level, thus maintaining the desirable noise immunity and extending the range for the transmission of the signal.

Each of the gates discussed so far has a corresponding negated version: the AND, OR, XOR, and buffer gates have the NAND, NOR, XNOR, and inverter gates as complements. The truth tables for these negated gates are the same as for the original gates except the output states are reversed. Thus the output states 0,0,0,1 for the AND gate become 1,1,1,0. The circuit symbol is the same except for a small circle on the output which indicates the inversion of the level. Finally, the Boolean symbol is changed by placing a bar over the original expression: thus $A \cdot B$ becomes $\overline{A \cdot B}$, and so on. These negated gates are shown in Figs. 8.8, 8.9, 8.10, and 8.11.

Figure 8.10 Circuit symbol, algebraic expression, and truth table for the XNOR gate.

A	B	Out
0	0	1
1	0	0
0	1	0
1	1	1

$$\text{Out} \equiv \overline{A \oplus B}$$

Figure 8.11 Circuit symbol, algebraic expression, and truth table for the inverter gate.

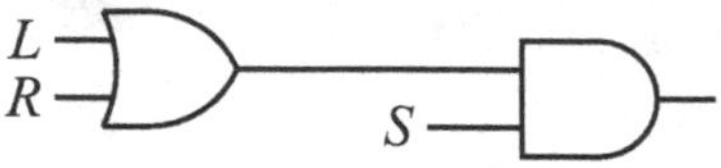

A	Out
0	1
1	0

$$\text{Out} \equiv \overline{A}$$

$$\text{Out} = (L + R) \cdot S$$

Figure 8.12 Solution of the car alarm problem.

8.5 Implementing logical functions

The implementation of simple logical functions can usually be determined after a little thought. For example, suppose you are designing a safety system for a two-door car. You want to sound an alarm (activated by a high level) when either door is ajar (this condition being indicated by a high logic level), but only if the driver is seated (again, indicated by a high level). Such a logic function is produced by the circuit in Fig. 8.12. The state of the left and right doors is represented by inputs L and R, while input S tells the circuit if the driver is seated. Thus if L or R is high (or both), and S is high, the output is high and the alarm sounds, as required.

With more complicated logic problems, the solution is less obvious. For such problems the *Karnaugh map* provides a method of solution. This method works for logic circuits having either three or four inputs. The first step in the method is to make a truth table for the problem. This follows from analyzing the requirements of our problem: under what conditions do we require a high output? As an example, suppose our analysis gives us the truth table shown in Fig. 8.13. For this example, we have three inputs, A, B, and C giving the output levels indicated.

The next step is to construct a Karnaugh map from the data in our truth table. This is illustrated in Fig. 8.14. The input states are listed along the top and left side of the map. For this example, with three inputs, we list the possible AB combinations along the top and the two C states along the left side.[4] When we

---

[4] For four inputs, the possible CD combinations would be listed along the left side as in Fig. 8.16.

A	B	C	Out
0	0	0	0
0	0	1	0
0	1	0	0
0	1	1	1
1	0	0	0
1	0	1	1
1	1	0	0
1	1	1	1

Figure 8.13 Truth table for the Karnaugh map example.

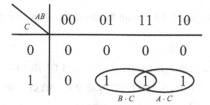

Figure 8.14 The Karnaugh map corresponding to Fig. 8.13.

list the AB combinations, we must follow a convention: only one digit at a time is changed as we write down the various combinations. In this example, we start (arbitrarily) with 00, and then change the second digit to get 01. To get another combination not yet listed, we change the first digit and get 11, and finally change the second digit obtaining 10. We then fill in the map with the data from the truth table.

The final steps are to identify groups of ones and then read the required logic from the map. The rule is to look for horizontal and/or vertical groups of 2, 4, 8, or 16. Diagonal groups are not allowed. In our example, there are two groups each containing two members. These are circled in Fig. 8.14. Now we identify the logic describing each group. To be a member of the group on the left, both B and C must be high, so the logic is $B \cdot C$. To be in the group on the right, both A and C must be high, so the logic is $A \cdot C$. Since a high output is obtained if we are a member of either group, the full logic describing our truth table is $(B \cdot C) + (A \cdot C)$. The implementation of this is shown in Fig. 8.15.

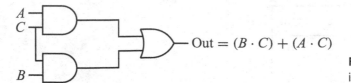

Out $= (B \cdot C) + (A \cdot C)$

Figure 8.15 The logic circuit implementation of Fig. 8.13.

AB\CD	00	01	11	10
00	0	0	0	0
01	0	0	0	0
11	1	0	0	1
10	1	0	0	1

Figure 8.16 A Karnaugh map example showing how edges connect.

Out $= \overline{C \cdot \overline{B}}$

Figure 8.17 Circuit for the negated version of Fig. 8.16.

When looking for groups in the Karnaugh map, the edges of the map connect. This is illustrated in the map shown in Fig. 8.16. Because we can connect the right and left edges of the map, the ones in this map form a group of four, as indicated. To be a member of this group, C must be high and B must be low. Thus the logic for this group is $C \cdot \overline{B}$. This is much simpler logic than we would obtain if we instead identified two groups of two in our map.

Another simplification results in cases where the map has many ones and few zeros. In such cases, we can identify groups of zeros, find the logic for being a member of these groups, and apply an inversion to the result. For example, if the ones and zeros of the central portion of the map in Fig. 8.16 were reversed, we would find one group of four zeros. As we have seen, the logic for this group is $C \cdot \overline{B}$, but now we invert the result, obtaining $\overline{C \cdot \overline{B}}$. This final inversion could be done by using a NAND gate, as shown in Fig. 8.17

8.6 Boolean algebra

An *algebra* is a statement of rules for manipulating members of a set. You have, no doubt, learned in the past rules for doing mathematical manipulations with

Table 8.3 The Boolean algebra

Defining OR	$0 + A = A$
	$1 + A = 1$
	$A + A = A$
	$A + \overline{A} = 1$
Defining AND	$0 \cdot A = 0$
	$1 \cdot A = A$
	$A \cdot A = A$
	$A \cdot \overline{A} = 0$
Defining NOT	$\overline{\overline{A}} = A$
Commutation	$A + B = B + A$
	$A \cdot B = B \cdot A$
Association	$A + (B + C) = (A + B) + C$
	$A \cdot (B \cdot C) = (A \cdot B) \cdot C$
Distribution	$A \cdot (B + C) = (A \cdot B) + (A \cdot C)$
	$A + (B \cdot C) = (A + B) \cdot (A + C)$
Absorption	$A + (A \cdot B) = A$
	$A \cdot (A + B) = A$
DeMorgan's 1	$\overline{A + B} = \overline{A} \cdot \overline{B}$
DeMorgan's 2	$\overline{A \cdot B} = \overline{A} + \overline{B}$

integers, real numbers, and complex numbers. There is also a special algebra for logical operations. It is called *Boolean algebra.*

The rules for Boolean algebra are shown in Table 8.3. They consist of definitions for the AND, OR, and NOT (or inversion) operations, and several theorems. In the table, A, B, and C are logical variables that can have values of 0 or 1. Once the definitions are accepted, the theorems can be proved by brute force by plugging in all the possible cases; since the variables have only two values, this is not too trying.

Boolean algebra can be used to find alternative ways of expressing a logical function. Consider the XOR function defined in Fig. 8.6. To get a high output, this function requires either A high while B is low, or B high while A is low. In algebraic terms,

$$A \oplus B = (A \cdot \overline{B}) + (B \cdot \overline{A}). \tag{8.3}$$

This equation shows us a way of producing the exclusive-OR function (other than buying an XOR gate). The resulting circuit is shown in Fig. 8.18. Note that in this figure (as in other figures in this chapter) we use the convention that crossing lines are not connected unless a dot is shown at the intersection point. This allows for more compact circuit drawings.

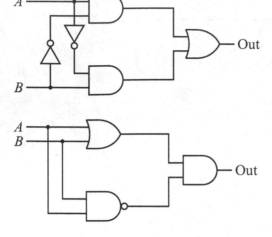

Figure 8.18 An alternative way of making an XOR gate.

Figure 8.19 Another way of making an XOR gate.

Now we employ some algebraic manipulations to find another (and simpler) way to express the XOR function. In the first line of Eq. (8.4), we use the fact that $A \cdot \overline{A} = 0$ and $0 + A = A$ (for any A) to rewrite Eq. (8.3). The next line uses the Distribution Theorem to group terms together. The third line uses the second DeMorgan Theorem and the last line again uses the Distribution Theorem. The resulting logic is implemented in Fig. 8.19. Note that this way of making an XOR gate is simpler than that in Fig. 8.18 because it uses fewer gates:

$$A \oplus B = (A \cdot \overline{B}) + (B \cdot \overline{A}) + (A \cdot \overline{A}) + (B \cdot \overline{B})$$
$$= A \cdot (\overline{A} + \overline{B}) + B \cdot (\overline{A} + \overline{B})$$
$$= A \cdot \overline{(A \cdot B)} + B \cdot \overline{(A \cdot B)}$$
$$= (A + B) \cdot \overline{(A \cdot B)}. \tag{8.4}$$

We have seen that we can construct an XOR gate from combinations of other gates. There is an interesting theorem that states that *any* logic function can be constructed from NOR gates alone, or from NAND gates alone. For example, suppose we want to make an AND gate from NOR gates. Using Boolean algebra, we can find the way:

$$A \cdot B = \overline{(\overline{A} + \overline{B})} = \overline{\overline{(A + 0)} + \overline{(B + 0)}}. \tag{8.5}$$

In the first equality, we have used the second DeMorgan Theorem and in the second equality we have used the fact that anything OR'd with 0 remains the same. The point is that the final expression is all in terms of NOR functions. The resulting circuit is shown in Fig. 8.20.

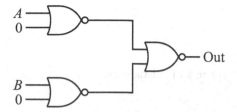

Figure 8.20 Making an AND gate from NORs.

It may seem that this is a silly thing to do. If you need an AND, why not just buy an AND instead of making it from NORs? There are two reasons. The first concerns the way logic gates are packaged. A typical integrated circuit (IC) chip will have four or six gates on a single chip, but all the gates are the same type (e.g., all NORs). Now if you are building a logic circuit that needs one NOR gate and one AND gate, you can buy two integrated circuits (one with NOR gates on it and one with AND gates on it) *or* you can use a single NOR gate IC containing at least four gates. In the latter case, one of the gates is used for the NOR function and the other three are used, as in Fig. 8.20, to produce the AND function. Thus you have saved money and circuit board space by using the NOR equivalent for the AND.

The second reason is, again, a practical one. If one is working on a logic circuit and runs out of one type of gate, is it useful to know that you can make do with a combination of NOR or NAND gates. Alternatively, if you are stocking an electronic workshop, you could just buy NOR or NAND gates instead of stocking all the different logic gates; you could always construct a needed function from the one type of gate you had on hand.

8.7 Making logic gates

Although we have discussed how logic gates function, we have not yet indicated how to make them. There are, in fact, many ways to make logic gates. A simple, low-tech way is to use an electromagnetic switch or *relay*, as shown in Fig. 8.21. The relay has a solenoid with a movable iron core that is mechanically attached to a switch. When a voltage is applied to the control input, the iron core is pulled into the solenoid and closes the switch. Without a control voltage, a spring (not shown) returns the switch to an open position.

Figures 8.22 and 8.23 show the use of relays to form an AND gate and an OR gate. The gate inputs A and B are connected to the relay controls and close the relevant switch when they are high. For the AND gate, two relays are connected in series, and for the OR gate, two relays are connected in parallel. When the switches

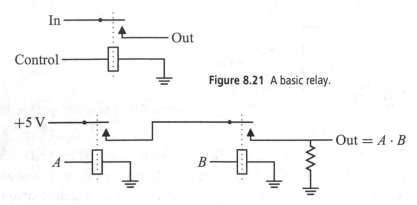

Figure 8.21 A basic relay.

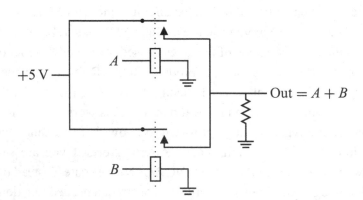

Figure 8.22 AND gate made from relays.

Figure 8.23 OR gate made from relays.

are open, the output is held at ground potential by the resistor R. When the logic is satisfied, the output is connected to the $+5$ V supply voltage and is thus high.

Semiconductor logic gates come in various types or families. One common type is the transistor-transistor logic (or *TTL*) family. Here bipolar transistors are used to create the logic gates. For example, the TTL NOR gate is shown in Fig. 8.24. If either A nor B is high, its transistor is driven into saturation, making the collector-emitter voltage small and the gate output low. If neither A nor B is high, both transistors are off, so there is no voltage drop across R and the output is high $(+5$ V$)$.

Another important logic family is the complementary metal oxide semiconductor (or *CMOS*) family. These circuits employ field-effect transistors. For example, a CMOS NOT gate is shown in Fig. 8.25. When a high voltage is applied to the input A, the transistor turns on and the voltage across it becomes small, thus giving a low output. When a low voltage is applied, the transistor does not conduct, so the output remains at $+5$ V (i.e., high).

Table 8.4 Characteristics of some logic families

Family	Pros	Cons
TTL	Common, fast, cheap	High power consumption
CMOS	Low power consumption, suitable for large scale integration	Relatively slow
ECL	Fastest	High power consumption, low noise immunity

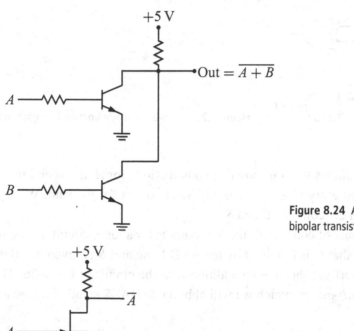

Figure 8.24 A TTL NOR gate made with bipolar transistors.

Figure 8.25 A CMOS NOT gate made with a field-effect transistor.

There are also special purpose logic families, like the fast-switching ECL (emitter coupled logic) family, but we will leave these for more advanced study. Some of the pros and cons of the families we have mentioned are shown in Table 8.4.

8.8 Adders

In addition to performing logic functions, gates can also be used to *add* binary numbers. To see this, consider the data in Table 8.5. We imagine that A and B are two binary numbers consisting of one bit, so each can only have the value 0 or 1.

Table 8.5 Adding two one-bit binary numbers

A	B	Sum	C	S
0	0	0	0	0
1	0	1	0	1
0	1	1	0	1
1	1	10	1	0

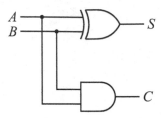

Figure 8.26 A half-adder made from an AND gate and an OR gate.

Our table shows the four possible combinations for A and B and their sum. Note that the last entry $(1 + 1 = 10)$ requires two bits for the sum. We separate these two bits in the columns C and S.

If we think of columns C and S as outputs for a logic circuit having inputs A and B, we see that C is provided by the AND logic and S is given by XOR logic. We can thus perform this one-bit addition with the circuit of Fig. 8.26. This circuit is called a *half-adder*, which we will abbreviate HA. S stands for *sum* and C stands for *carry*.

Adding one-bit binary numbers is nice, but usually our arithmetic needs will involve longer binary numbers. Perhaps we can use a series of n half-adders to add an n-bit binary number. Unfortunately, this does not work. If we review the binary addition example in Table 8.2, we see that each column of the addition requires *three* possible inputs: one from each number and one to accommodate the possibility of a carry from the sum of the digits to the right. The half-adder *outputs* a sum and a carry but has no provision for a carry *input*.

The solution to this problem, shown in Fig. 8.27, uses two half-adders and an OR gate. This new circuit is called a *full adder* (FA). The full adder has three inputs, thus allowing for a carry input from the result of a previous addition. The output has a carry and a sum, just as for the half-adder. The reader can verify the proper operation of this device by working through the logic for all possible input combinations.

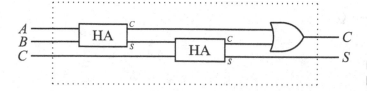

Figure 8.27 A full-adder made from two half-adders and an OR gate.

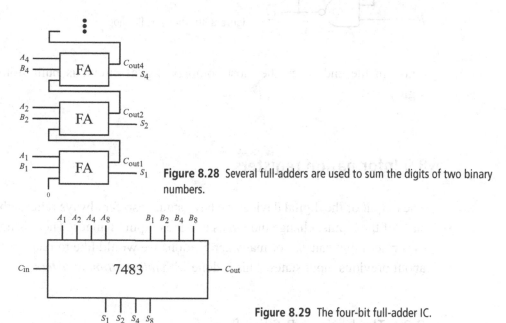

Figure 8.28 Several full-adders are used to sum the digits of two binary numbers.

Figure 8.29 The four-bit full-adder IC.

The use of full adders to add two four-bit binary numbers is illustrated in Fig. 8.28. We imagine that our first number has digits A_1, A_2, A_4, and A_8, and that our second number has digits B_1, B_2, B_4, and B_8 (the subscripts are 2^n). In our circuit, each of these digits is represented by a logic level on a separate line. These are connected to a series of full adders as shown, with the carry out of one full adder connected to the carry in of the next. The carry input of the first full adder is not used and is thus connected to a low level. The sum output of each adder becomes a digit (S_1, S_2, S_4, and S_8) of the resulting sum, with the carry out from the last full adder giving S_{16}.

A four-bit full adder is available as an integrated circuit, shown schematically in Fig. 8.29. It has inputs for the four bits of A and B and outputs the four bits of S as well as a possible additional bit (the carry out). The carry in connection allows for the connection of multiple units when adding a number with more than four bits.

It is worth noting at this point how the complexity of our circuit has multiplied. Each half-adder uses two gates, so a full-adder has five gates, and the four-bit adder has twenty gates. Yet the usage rules for the four-bit adder are relatively simple. This is characteristic of digital circuitry: we build functionality by combining smaller

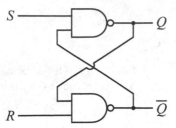

Figure 8.30 The basic flip-flop.

units. In the end, even the most complex digital circuit is built from humble logic gates.

8.9 Information registers

The output of the digital devices we have studied so far always reflects the current state of the inputs: change the inputs and the output changes. There is no memory of former input states. For many applications we would like to retain information about previous input states. This is done with *information registers*.

8.9.1 The basic or R-S flip-flop

The simplest information register is shown in Fig. 8.30. It is called by various names: the *basic flip-flop, binary*, or the *R-S flip-flop* (RSFF). The circuit has inputs S and R (which stand for Set and Reset) and outputs labeled Q and $\overline{Q}$. We emphasize that Q and $\overline{Q}$ are just labels. In most cases these will be in opposite logic states, but not always.

To see how this circuit functions, suppose that both S and R are high. What will be the state of Q and $\overline{Q}$? Because of the feedback between the outputs and the NAND gate inputs, we have to look for states that are self-consistent (i.e., that satisfy the logic of the entire circuit). Suppose, for example, that we assume $Q=1$ and $\overline{Q}=0$. If we follow the circuit, putting $Q=1$ into the bottom NAND gate along with $R=1$, we get a zero output, which is consistent with our assumption of $\overline{Q}=0$. Putting $\overline{Q}=0$ into the top gate along with $S=1$ gives a output of 1, which is consistent with our assumption for Q. Thus the output state $Q, \overline{Q} = 1, 0$ satisfies the logic of the circuit. The reader can verify that the state $Q, \overline{Q} = 0, 1$ also satisfies the logic of the circuit, while $Q, \overline{Q} = 0, 0$ and $1, 1$ do not. Thus this circuit has two stable output states, or is *bistable*.

Table 8.6 Response of the basic flip-flop to changes in inputs

Case	Time	R	S	Q	$\overline{Q}$
1	Start	1	1	0	1
	End	1	0	1	0
2	Start	1	1	1	0
	End	1	0	1	0
3	Start	1	1	1	0
	End	0	1	0	1
4	Start	1	1	0	1
	End	0	1	0	1

The point of the circuit is that we can choose which of these two states the circuit will be in by making either R or S momentarily zero. The four possible cases are summarized in Table 8.6. For each case we start with both R and S high and Q and $\overline{Q}$ in one of the two stable states. Then we examine what happens when we make R or S low. In case 1, we see that if $Q, \overline{Q} = 0, 1$ initially, then making S low changes the outputs to $Q, \overline{Q} = 1, 0$. Case 2 shows that if $Q, \overline{Q} = 1, 0$ initially, then making S low has no effect. Similarly, cases 3 and 4 show that making R low leaves the outputs in the state $Q, \overline{Q} = 0, 1$ regardless of the initial state.

We can summarize these results by saying that the current output state tells us which input was low last. If S was low last, then the output state will be $Q, \overline{Q} = 1, 0$. If R was low last, the output state will be $Q, \overline{Q} = 0, 1$. In shorthand, changes are governed by

$$(R, S) \rightarrow (Q, \overline{Q}) \tag{8.6}$$

assuming only one of the inputs is low.[5]

8.9.2 The clocked flip-flop

While the basic flip-flop remembers which input was low last, this memory is limited; the next time R or S changes, the former result is lost. This is the motivation for our next circuit, shown in Fig. 8.31. This circuit is called the *clocked* or *gated R-S flip-flop*. The right part of the circuit is the basic flip-flop we studied in the

---

[5] If both R and S are low, the outputs become $Q, \overline{Q} = 1, 1$. But when R and S are returned to their normal high state, this output cannot remain since it is not a stable state. The output falls into one of the stable states, but we cannot predict which one.

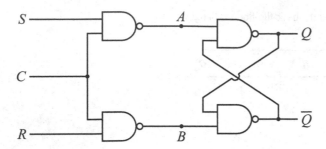

Figure 8.31 The clocked flip-flop.

last section; recall that this requires a low input to change states. The inputs to this basic flip-flop come from the NAND gates on the left.

The truth table for a NAND gate tells us that it will only have a low output if both inputs are high. Our clocked flip-flop thus works as follows: we imagine that R and S are normally low and that occasionally one of them becomes high. If the clock input C is also high, the corresponding NAND gate can output a low and thus change Q and $\overline{Q}$. If C is low, however, changes to R and S have no effect on Q and $\overline{Q}$. In shorthand, changes are governed by

$$(S, R) \rightarrow (Q, \overline{Q}) \tag{8.7}$$

but only if C is high and assuming only one of the inputs S, R is high.

Our "clock" input C thus provides a degree of isolation between the inputs S and R and the outputs Q and $\overline{Q}$. The input information is only "read" and "saved" in the output state when C is high. Our memory is now more selective and permanent: we only store input information when C is high and we can keep that information expressed on the outputs as long as C is low.

8.9.3 The M-S flip-flop

The clocked flip-flop of the last section isolates changes in the inputs from changes in the outputs through the use of the clock input. The output can change only when C is high. If, however, there are *multiple changes* in the levels of the inputs S and R during the time C is high, the outputs will also change multiple times. We would like to take the process of isolating input changes from output changes one step further so that the state of the inputs S and R at one instant of time will be read and saved.

We can accomplish this using the circuit shown in Fig. 8.32. Two clocked flip-flops (CFF) are used along with a NOT gate to form the *master-slave or M-S flip-flop* (MSFF). When C is low, changes in the inputs S and R will not affect the

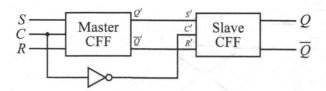

Figure 8.32 The M-S flip-flop.

outputs of the master flip-flop Q' and $\overline{Q}'$. The clock input C' of the slave flip-flop, however, is high, and its inputs S' and R' are directly connected to Q' and $\overline{Q}'$. Thus the current state of Q' and $\overline{Q}'$ is reflected in the final outputs Q and $\overline{Q}$.

When C is high, changes in S and R are reflected in the master outputs Q' and $\overline{Q}'$, but since C' is low, these do not change the final outputs Q and $\overline{Q}$. When C changes from high to low, the state of S and R at that instant will also be present at the input of the second flip-flop, and this input will be read as C' changes from low to high. We have thus created a circuit that reads and stores the input states S and R only at the high-low transition of the clock input C. This is called *negative transition edge clocking*. Again, in shorthand, changes are governed by

$$(S, R) \rightarrow (Q, \overline{Q}) \tag{8.8}$$

but only at the high-low transition of C and assuming only one of the inputs S,R is high.

8.9.4 Other flip-flop variations

There are other variations in flip-flop construction and operation that should be noted. The clocked flip-flop and M-S flip-flop can also be constructed from NOR gates rather than NANDs. The result is a CFF that changes when the clock is low rather than high, and an MSFF that changes on the low-high transition of the clock rather than the high-low transition.

The simple summaries of our flip-flop operations given by Eqs. (8.6), (8.7), and (8.8) are accompanied by assumptions about the logic levels on the S and R inputs. If both inputs are high for the CFF or MSFF (or both low for the RSFF), the output state becomes unpredictable when the inputs return to their normal state. One way to deal with this problem is to add feedback lines from the outputs Q and $\overline{Q}$ to the inputs. A clocked flip-flop with this modification is shown in Fig. 8.33. Following common usage, the S and R inputs are renamed J and K, respectively, and the circuit is called a *clocked J-K flip-flop*. The reader can verify that this circuit follows the rules of the clocked RSFF with the following change. If both J and K are high when the clock goes high, the outputs switch (or toggle) from whichever

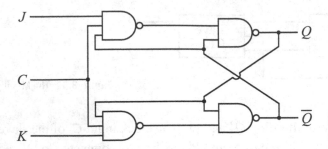

Figure 8.33 The *J-K* modification of the clocked flip-flop.

state they were in to the opposite state. Thus, $Q, \overline{Q} = (1, 0)$ becomes $(0, 1)$ and vice versa. Similarly, a *J-K* version of the MSFF can be constructed.

Finally, we note that direct set and reset (or clear) inputs are often added to flip-flops. These provide an additional way to change the outputs that usually overrides the other rules of operation. Thus, a prescribed logic level (this might be high or low depending on the device details) applied to the direct set input will immediately produce $Q, \overline{Q} = (1, 0)$ regardless of what the clock and other inputs are doing. Conversely, when the prescribed level is applied to the direct reset input, $Q, \overline{Q} = (0, 1)$ is immediately produced.

8.10 Counters

We have seen that flip-flops can be used to store information about the logic levels of their inputs. They are also used to produce other useful functions. One example of this is shown in Fig. 8.34 where the outputs of an MS flip-flop are connected to the inputs. The result is a *toggle flip-flop* (TFF). To see how this works, we imagine that the clock input C is regularly switching states, as shown in Fig. 8.35. The outputs Q and $\overline{Q}$ will be in opposite states, and we suppose they start out as $Q, \overline{Q} = (0, 1)$. The outputs will change to reflect the input states at the high-low transition of the clock C, in accordance to Eq. (8.8). Since $\overline{Q}$ is connected to S in our circuit, the state of $\overline{Q}$ at the transition will end up at Q. Similarly, the state of Q at the transition will end up at $\overline{Q}$. The $(Q, \overline{Q})$ states thus switch or *toggle* at the negative transition edge of C. If the clock waveform is a square wave as in Fig. 8.35, then Q and $\overline{Q}$ are also square waves, but with twice the period of C, or half the frequency. Because of this latter interpretation, the toggle flip-flop is sometimes called a *divide-by-two*.

If we connect several toggle flip-flops as shown in Fig. 8.36, we obtain a *binary counter*. The output of the leftmost TFF is used as the clock for the next TFF, and this connection strategy is repeated as we move to the right. The $\overline{Q}$ outputs are

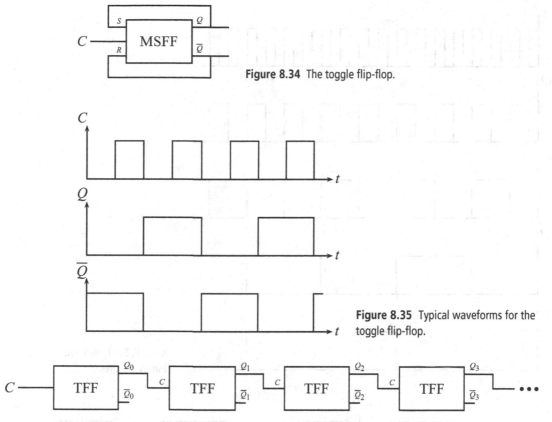

Figure 8.34 The toggle flip-flop.

Figure 8.35 Typical waveforms for the toggle flip-flop.

Figure 8.36 Toggle flip-flops joined to form a binary counter.

not used. The output of each TFF changes state as its particular clock waveform undergoes a high-low transition. The output waveforms for the first four TFFs are shown in Fig. 8.37.

To see why this circuit acts as a counter, think of the outputs Q_0, Q_1, Q_2, and Q_3 as the digits of a binary number $Q_3 Q_2 Q_1 Q_0$ that gives the number of high-low transitions in C. We start with all the outputs low, so the count is 0000_2. After the first high-low transition, Q_0 is high, so we have 0001_2, as shown by the leftmost dotted line in Fig. 8.37. After seven high-low transitions, Q_0, Q_1, and Q_2 are high and our number is 0111_2, as shown by the next dotted line. One more transition gives 1000_2, matching the eight transitions that have occurred (see third dotted line). This process continues until the count reaches fifteen (1111_2). The next transition gives 0000_2 and our counter has reset to its initial state (last dotted line). Clearly, this can be extended to use any number n of toggle flip-flops, in which case the counter will reset to zero after 2^n clock transitions.

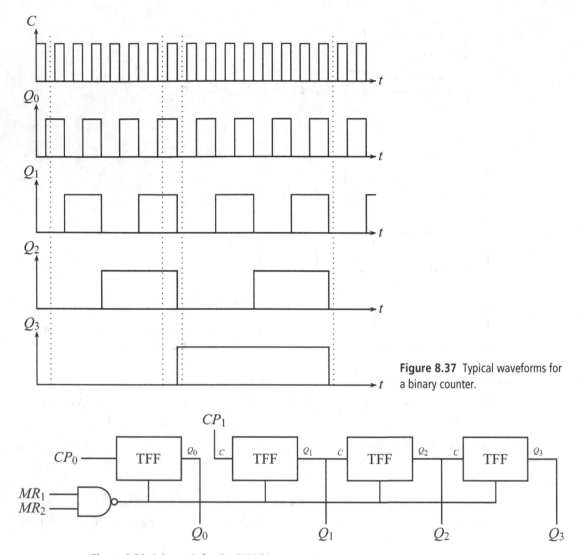

Figure 8.37 Typical waveforms for a binary counter.

Figure 8.38 Schematic for the 7493 binary counter.

Counting circuits are available as integrated circuits. The functional schematic for one such IC (the 7493) is shown in Fig. 8.38. For flexibility, one TFF is independent, with its own clock input CP_0. The other three TFFs are connected internally as needed for a binary counter and are driven by CP_1. To make a four-bit counter, we simply connect Q_0 to CP_1.

This circuit has an additional useful feature. There are two *master reset* inputs MR_1 and MR_2. When both of these inputs are high, the counter immediately resets all outputs to zero. By connecting one or two of the outputs to these master

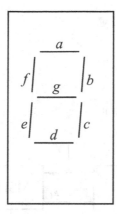

Figure 8.39 A seven-segment display showing the segment labels *a–g.*

resets, we can force the counter to return to zero after reaching a certain count. For example, if we connect Q_2 and Q_0 to MR_1 and MR_2, the first time these two outputs are high (i.e, 0101_2), the counter will immediately return to zero. Thus the count will go like this: 0000, 0001, 0010, 0011, 0100, 0000, etc. A counter that returns to zero after n counts is called a *modulo-n counter*. Using this terminology, in this example we have taken a modulo-16 counter and made it into a modulo-5 counter by employing the master resets. Since our usual base 10 number system resets after 10 counts (0,1,2,...9,0,...), a modulo-10 counter is given a special name: a *binary-coded decimal* or *BCD* counter.

8.11 Displays and decoders

The usual number displays in our world are decimal, using the digits zero through nine. A common device used for such numbers is the *seven-segment display*, shown in Fig. 8.39. The device has seven independent line segments (*a* through *g*) which can be lit by applying a high level to the appropriate input. By lighting the appropriate segments, a boxy version of each of the ten decimal digits can be produced.

As we have seen, our counter circuits produce binary representations of numbers. If we want to display these numbers using a seven-segment display, we need a circuit that will take the four outputs of our BCD counter and light up the appropriate segments of the display. Such a device is called a *decoder*, and a typical IC is shown in Fig. 8.40. It has inputs for the BCD digits Q_0, Q_1, Q_2, and Q_3 and outputs for each of the segments of the seven-segment display.

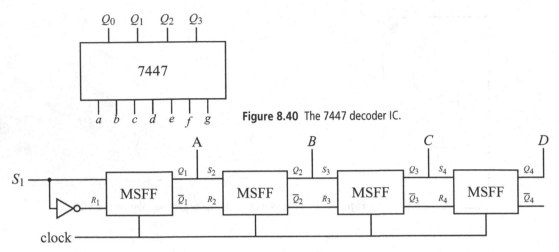

Figure 8.40 The 7447 decoder IC.

Figure 8.41 Using MSFFs to make a shift register.

8.12 Shift registers

Another way to employ MS flip-flops is to join them together to form a *shift register*, as shown in Fig. 8.41. In this case, all the flip-flops have a common clock input which is typically a square wave as shown in Fig. 8.42. A series of high and low levels is applied to S_1 synchronized with the clock. Since the MSFF operation follows Eq. (8.8), the level present at S_1 at the high-to-low transition of the clock will be transfered to Q_1. At the next transition, this will be transfered to Q_2, and so on. The net effect is that the waveform sequence applied at S_1 is transferred down the line of flip-flops, one step for each negative transition edge of the clock. If we wish to repeat the pattern, the output of the last flip-flop can be returned to the input of the first.

8.12.1 Shift register applications

Shift registers have a number of applications and we mention two of them here. The first is the production of a scrolling message sign of the type often seen in public places. If we connect each of the outputs *A, B, C...,* of Fig. 8.41 to an LED in a long row of LEDs, we would see our input pattern move down the row at a rate set by the clock. If we do the same thing with a number of shift register/LED rows and arrange the rows one under the other, we obtain a rectangular array of LEDs controlled by the various flip-flops in the shift registers. We can now load any pattern we choose into the array and propagate it down the line. Often this pattern

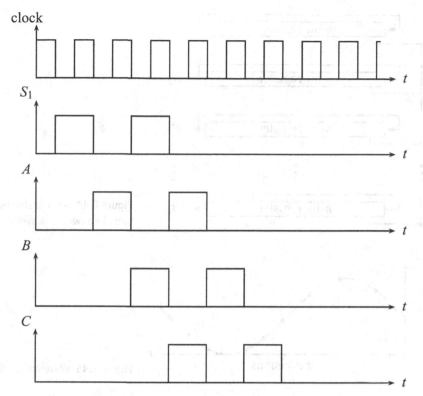

Figure 8.42 Example waveforms for the shift register.

Figure 8.43 Scrolling message display.

is chosen to form the letters of a message as in Fig. 8.43. The word "HELLO" would move one column to the right with each clock pulse.

We can also use multiple shift registers for *digital waveform synthesis*. For this application we imagine we have m shift registers, each of which has n flip-flops and thus can hold n logic levels. We use only the output of the last flip-flop of each shift register. At any given time the logic levels of these m outputs form an m-bit binary number B. Each time the clock pulses, a different number B appears on the outputs, and, since each shift register has n flip-flops, we can produce n different m-bit numbers. The output of the last flip-flop of each shift register is connected to the input of the first flip-flop so that the pattern repeats. This setup is shown in Fig. 8.44.

Now suppose we wish to create a periodic waveform. We divide the period of the waveform into n equal divisions as shown in Fig. 8.45. An m-bit binary number

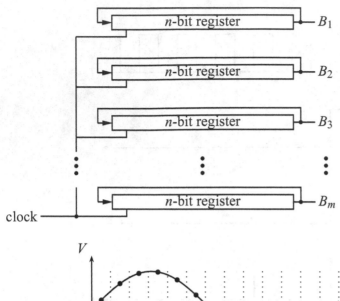

Figure 8.44 A set of shift registers used to make a waveform synthesizer.

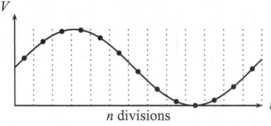

n divisions

Figure 8.45 Waveform synthesizer wave.

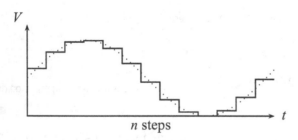

n steps

Figure 8.46 Synthesized wave (solid line). The dotted line shows the waveform after low-pass filter.

proportional to the amplitude of the waveform at each time is then loaded into the shift register array. After all this information is loaded, we run the clock at its normal rate. The circuit now gives a series of binary numbers B at the output that represent the amplitude of the waveform at subsequent time steps. If we can translate these numbers into an analog voltage (see Section 8.13), we can produce a stepped version of the original waveform as in Fig. 8.46. This can then be passed through a low-pass filter which will smooth it into a sine wave.

Although this method of producing waveforms seems more complicated than using an analog circuit, it has certain advantages. The most striking is that we can reproduce *any periodic waveform* with this method. Our analog circuits can produce certain waveforms (sine, sawtooth, square waves), but would be hard pressed to make a more irregular pattern. Second, the frequency of our synthesized waveform is easily changed by changing the clock rate. Lastly, the circuit design is easily extended; we can improve the precision (in time or amplitude) of our synthesized waveform by increasing the number of shift registers or the number of flip-flops in each shift register.

8.13 Digital to analog converters

The last step in our digital waveform synthesizer is to take the series of binary numbers B presented at the output of our array of shift registers and convert them to analog voltages. This requires a device called a *digital to analog converter* (also written as *D/A converter* or *DAC*). It has an input for each bit of the binary number and outputs a voltage proportional to that number. There are various ways to do this, but a simple one which uses our previous knowledge is shown in Fig. 8.47. It uses the op-amp adder circuit developed in Section 6.3. Recall that this circuit produces a *weighted sum*, with the weights set by the resistor values:

$$V_{\text{out}} = -\left(\frac{R_f}{R_1}V_1 + \frac{R_f}{R_2}V_2 + \frac{R_f}{R_3}V_3 + \frac{R_f}{R_4}V_4\right). \tag{8.9}$$

Since our input voltages come from a digital circuit, all high levels will be at the same voltage and similarly for the low levels. If these represent the bits of a binary number, however, we want the more significant bits to count more than those of lesser significance. This can be achieved by weighting the sum. If we choose the resistor values shown in Fig. 8.47, Eq. (8.9) reduces to

$$V_{\text{out}} = -(V_1 + 2V_2 + 4V_3 + 8V_4). \tag{8.10}$$

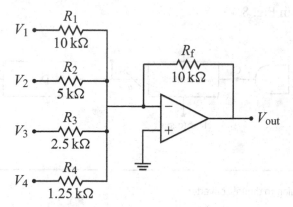

Figure 8.47 A four-bit digital to analog converter.

Thus, if we connect the 2^0 bit of our binary number to V_1, the 2^1 bit to V_2, the 2^2 bit to V_3, and the the 2^4 bit to V_4 we will get an output voltage that is proportional to the binary input number. Clearly, this scheme can be extended to any number of bits.

In most cases, a user will simply buy a D/A converter on a chip rather than build one from components. In addition to our scaled resistor method, there are several other ways to achieve digital to analog conversion, and the interested reader should consult the end-of-chapter references for details. Here we briefly discuss some of the issues common to all DACs.

The *resolution* of a D/A converter refers to the smallest change step at the output. This is set by the number of input bits. For an *n*-bit input, the resolution is one part in 2^n. *Linearity* is a measure of how much the output varies from a perfect proportionality to the input binary number. The *accuracy* tells you how well the proportionality matches a specified value. For our op-amp example, linearity and accuracy would depend on how well the resistors matched the desired values. Finally, the *settling time* is the time required for the output to get within some specified amount of its final value. For the op-amp DAC, the settling time would depend on the slew rate of the op-amp.

8.14 Analog to digital converters

While most computation and data analysis is done on digital computers, most laboratory signals are analog. We therefore need a device that will change our lab signals into binary numbers, and this device is called an *analog to digital converter* (also written as *A/D converter* or *ADC*). Again, there are several different methods to achieve this. Here we discuss a simple method that uses some of our previously studied devices. The schematic for a *staircase A/D* is shown in Fig. 8.48 with associated waveforms in Fig. 8.49.

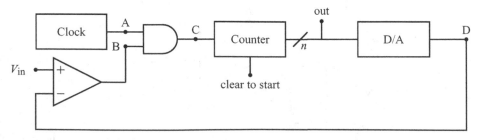

Figure 8.48 A staircase analog to digital converter.

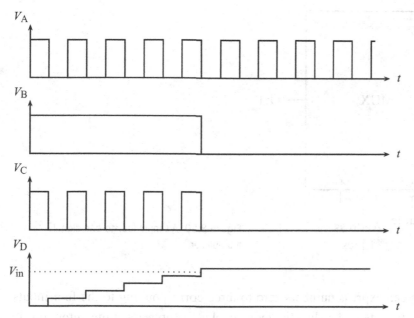

Figure 8.49 Waveforms for the staircase analog to digital converter.

The voltage to be converted, V_{in}, which we take to be positive, is applied to the (+) input of a comparator. To start the conversion, the binary counter is cleared (set to zero). All n output lines of the counter (represented by a single slashed line in Fig. 8.48) are thus low. The D/A converter then converts this zero binary number to zero voltage, which is fed into the (−) input of the comparator. Since $V_{in} > 0$, the comparator output is high. This allows the clock signal to pass through the AND gate and be counted by the counter. As the count increases, the output voltage, V_D, of the D/A converter increases in a staircase fashion. Eventually, V_D becomes greater than V_{in} and this causes the comparator output to go low. This in turn prevents the clock signal from passing through the AND gate so the counter stops. At this point the binary number at the output of the counter is proportional to the input voltage, and the conversion is complete.

8.15 Multiplexers and demultiplexers

Multiplexers and demultiplexers are devices that route signals. A *multiplexer* takes one of several inputs and connects it to a single output. The input line that is connected is determined by the binary number applied to the address lines. A schematic of a multiplexer with four input lines is shown in Fig. 8.50. The two

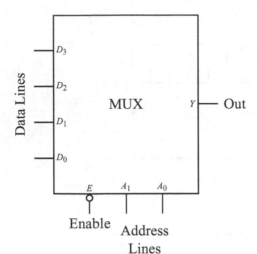

Figure 8.50 A simple four-to-one digital multiplexer.

address lines can express numbers zero to three corresponding to the four inputs. For example, if $A_1A_0 = 10$, then the logic level at D_2 appears at the output pin Y.

Several general comments about multiplexers can be made. The number of inputs on a multiplexer is not limited to four and is related to the number of address lines. It takes N address lines to support 2^N data lines. A digital multiplexer is unidirectional: the digital level at the selected data line is transferred to the output Y, but a digital level applied at Y will *not* be transfered to the selected data line. Finally, most multiplexers have an Enable input, as shown in Fig. 8.50. If this input is low, the multiplexer is enabled and the operation described above occurs. If Enable is high, the output is not connected to any data line and has a high impedance to ground. This allows one essentially to disconnect Y from whatever follows it. The small circle on the Enable input reflects the fact that the logic is reversed (a high level would normally be associated with turning a device on, not off).

As an application of the multiplexer, consider Fig. 8.51 which shows a parallel-to-serial converter. The parallel data presented at the four input lines are sequentially presented at the output as the counter driven by a clock oscillator steps through its four binary numbers: $A_1A_0 = 00$, 01, 10, and 11. After this the counter resets to zero and the process repeats. Although not shown, the Enable line could be employed to prevent the repetition.

A multiplexer can also be used to implement a truth table. A simple example of this is shown in Fig. 8.52. The truth table inputs A and B are connected to the address lines and the correct output level for the truth table is achieved by connecting the data input lines either high or low. Although this example would be trivial to

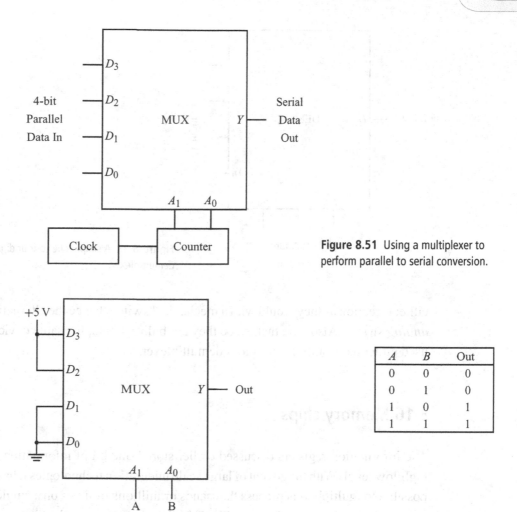

Figure 8.51 Using a multiplexer to perform parallel to serial conversion.

Figure 8.52 Using a multiplexer to implement a truth table.

implement without a multiplexer (Output = A), this method of implementing truth tables becomes more attractive (i.e., simpler and more compact) when the number of inputs increases.

The opposite of a multiplexer is a *demultiplexer*, shown in Fig. 8.53. In this case, the logic level at the single input D is routed to one of several output lines Y in accordance with the binary number applied to the address lines. Unused output lines have a high impedance to ground, as does the selected output if the Enable input is high. As with the multiplexer, digital demultiplexers are unidirectional.

Finally, we note that there exist analog versions of the multiplexer that will pass along an analog signal (i.e., one that can vary continuously) and not just high/low logic levels. These devices are typically bidirectional so that signals can pass in

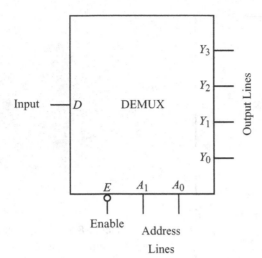

Figure 8.53 A simple one-to-four digital demultiplexer.

either direction as they would with a mechanical switch, hence the alternative name *analog switch*. Also note that, since they are bidirectional, the same device can be used either as a multiplexer or as a demultiplexer.

8.16 Memory chips

The information registers discussed earlier stored one bit of information, a single high/low level. With the advent of large scale integration technologies it has become possible to multiply this process thousands or millions of times on a single chip. A representative device is shown in Fig. 8.54. We first note that the bits are stored in groups called a data *word*. For our example, the word size is four bits, as seen by the four data lines. The number of words in the memory is limited to and typically set by the number of address lines. In our case, there are eight address lines and so $2^8 = 256$ words of memory on the chip. The Enable pin must be low to either read or write data. When it is high the data lines are essentially disconnected (have a high impedance to ground). When the Write pin is low, logic levels on the data lines are stored (written or input) to the memory word specified by the binary number on the address lines. When the Write pin is high, logic levels stored at the memory location specified by the address lines are placed on the data lines where they can be read by or output to other devices.

Memory chips come in a huge variety of sizes and types and carry their own set of acronyms and terminology. *Volatile* memory devices lose all stored information if the electrical power supply is removed, while *non-volatile* memory does not. *Static*

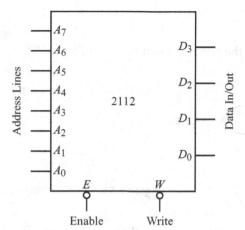

Figure 8.54 A 4-bit × 256 RAM.

memory stores data as long as the chip has power, while *dynamic* memory needs to have the stored information periodically refreshed or re-written to maintain storage. The example given above is an example of *RAM* or *random access memory* for which any data word can be accessed via the address lines in an order chosen by the user. In contrast, the shift register might be termed a sequential access memory because the bits have to be accessed in order of their position within the register.

The information in a *ROM* or *read only memory* cannot be routinely changed (written) but is intended as permanently stored information that will be read as needed. The stored information might be set at the time of manufacture, or, with a *PROM* or *programmable ROM*, set by the user using a special apparatus. With some PROMs, this step can occur only once, while an *EPROM* can be Erased and re-programmed.

EXERCISES

1. Develop the truth table for the circuit shown in Fig. 8.55.

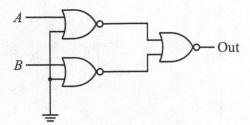

Figure 8.55 Circuit for Problem 1.

2. Using only NAND gates, construct a circuit that will implement the following
   logical expressions. Use Boolean algebra to simplify the expressions as much
   as possible before you begin.
   (a) $(A \cdot B) + (\overline{A} \cdot B) + (A \cdot \overline{B}) + (\overline{A} \cdot \overline{B})$
   (b) $[(A \cdot B) + C] \cdot [(A \cdot B) + D]$
   (c) $[(\overline{A} \cdot B) \cdot (A \cdot B)] + (A \cdot B)$
   (d) $(1 + B) \cdot (A \cdot B \cdot C)$

3. Using only NOR gates, give circuits that are equivalent to each of the following:
   AND, OR, NAND, and XOR.

4. Produce the truth table for the AND-OR-INVERT (AOI) gate shown in Fig. 8.56.

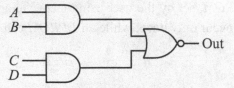

Figure 8.56 Circuit for Problem 4.

5. Without a calculator, find the binary equivalents of the following base 10
   numbers: 92, 66, 120, 511, 37, 255.

6. Using two-input logic gates, design an alarm system which lights an LED when
   any of five doors is open. Assume that an open door gives a high logic level.

7. Using only NOR gates, produce a clocked flip-flop with the same functionality
   as the one from Section 8.9.2. Use no more than five gates. Hint: try using the
   same configuration with the NAND gates replaced with NORs.

8. Suppose we want a four-person vote counter that will output a high level when
   three or four persons vote YES (indicated by a high input level). Produce a
   Karnaugh map for this problem and give the resulting logic circuit. You may
   use logic gates with any legal number of inputs.

FURTHER READING

Dennis Barnaal, *Digital and Microprocessor Electronics for Scientific Application* (Prospect Heights, IL: Waveland Press, 1982).

Don Lancaster, *TTL Cookbook* (Indianapolis, IN: Sams, 1983).

Don Lancaster, *CMOS Cookbook*, 2nd edition (Boston, MA: Newnes, 1997).

Sol Libes, *Small Computer Systems Handbook*, (Rochelle Park, NJ: Hayden, 1978).

Niklaus Wirth, *Digital Circuit Design for Computer Science Students: An Introductory Textbook* (New York: Springer, 1995).

Appendix A: Selected answers to exercises

Chapter 1

1. $1.27\ \Omega$

3a. $100\ W$

5. $I_2 = 0.397\ A$

7. $667\ \Omega$

9. $12.5\ \Omega$

11. $I_3 = 1.20A$

13. $V_{th} = 0.025\ V$, with positive terminal on the right

15. $V_2 = 15\ V$

17. $I_1 = -1.096\ A$, $I_3 = 0.896\ A$, $I_5 = 0.086\ A$

Chapter 2

1. $1.33\ \mu F$

3. $8.3\ \mu F$

6. magnitude $11.3\ k\Omega$, phase -27.95 degrees

8. $707\ rad/s$

10. $RC = 5.5 \times 10^{-5}\ s$

12. $|I| = 3.36\ A_{rms}$, phase -9.89 degrees

13. Turns ratio is $1/10$, primary current is $0.06\ A_{rms}$, secondary current is $0.6\ A_{rms}$

15. Output voltage will be $24\ V_{rms}$

16. The output is a 90 Hz sine wave with peak amplitude $0.946V_0$

Chapter 3

5a. If R_L is shorted, the power into the $\frac{1}{4}$ W resistor is 1.44 W, so the resistor burns out

6a. Circuit gives a constant V_{out} for $R_L > 3\ k\Omega$

8. Hint: my solution uses the LM317T voltage regulator

Chapter 4

2. $I_b = 90.4\ \mu A$, $I_c = 18.1\ mA$, $V_{ce} = 0.873\ V$

4. $r_{be} = 1250\ \Omega$, $\beta = 180$, $r_{out} = 5\ k\Omega$

7. $a = -3.97$, $g = -6.58$

9. $1.067\ V_{pp}$

Chapter 5

1c. $K = 1\ \text{mA/V}^2$, $V_t = -3\ \text{V}$

2. $V_{gs} = -1.35\ \text{V}$, $I_d = 2.71\ \text{mA}$, $V_{ds} = 4.16\ \text{V}$

4. $a = -1.77$, $g = -16.1$

Chapter 6

2. $V_{out} = V_3 + V_4 - V_1 - V_2$

3. $I = V_{in}/(3R_1)$

4. $R_{min} = 560\ \Omega$

7. $147\ \text{mV}$

8a. $V_{out} = -7.2\ \text{V}$

Chapter 7

1. Output is a sawtooth with amplitude 5.1 V and period 5.1 ms

5. $T = 2RC \ln \left(\dfrac{\frac{1}{3}(2V_{in} + V_{sat}^-) - V_{sat}^+}{\frac{1}{3}(2V_{in} + V_{sat}^+) - V_{sat}^+} \right)$

7a. $\omega = 1/RC$

Chapter 8

1. Truth table is the same as AND

2a. Equivalent to 1

2c. Equivalent to $A \cdot B$

5. $92_{10} = 1011100_2$, $66_{10} = 1000010_2$, $120_{10} = 1111000_2$

Appendix B: Solving a set of linear algebraic equations

B.1 Introduction

When analyzing a network of linear components (e.g., resistors, capacitors, and inductors), we typically obtain a set of linear algebraic equations for the unknown currents in the circuit. Cramer's Method, which is usually a topic in a linear algebra course, gives a method for solving such problems. The method can be applied for any number of unknown currents when we have an equal number of independent linear equations. For purposes of illustration, we will take the case where there are three unknown currents I_1, I_2, and I_3 related by three linear equations. Since the equations are linear, they can be cast in the form

$$a_{11}I_1 + a_{12}I_2 + a_{13}I_3 = b_1 \tag{B.1}$$

$$a_{21}I_1 + a_{22}I_2 + a_{23}I_3 = b_2 \tag{B.2}$$

$$a_{31}I_1 + a_{32}I_2 + a_{33}I_3 = b_3 \tag{B.3}$$

where the coefficients a_{ij} are known constants (real or complex) depending on the values of the circuit components, I_i are the unknown currents, and b_i are known constants (usually depending on the voltages in the circuit). Here i is the row index and j is the column index. This set of equations can be cast in matrix form as

$$\begin{pmatrix} a_{11} & a_{12} & a_{13} \\ a_{21} & a_{22} & a_{23} \\ a_{31} & a_{32} & a_{33} \end{pmatrix} \begin{pmatrix} I_1 \\ I_2 \\ I_3 \end{pmatrix} = \begin{pmatrix} b_1 \\ b_2 \\ b_3 \end{pmatrix}. \tag{B.4}$$

B.2 Cramer's Method

Cramer's Method is one way of obtaining the solution for the unknown currents I_i. A determinant D is formed from the coefficients a_{ij} of the unknown currents in Eqs. (B.1) through (B.3). The unknown currents I_i are found by forming this same determinant with the ith column replaced by the constants b_i and then dividing by D. For our case we obtain

$$D = \begin{vmatrix} a_{11} & a_{12} & a_{13} \\ a_{21} & a_{22} & a_{23} \\ a_{31} & a_{32} & a_{33} \end{vmatrix}, \tag{B.5}$$

$$I_1 = \frac{1}{D} \begin{vmatrix} b_1 & a_{12} & a_{13} \\ b_2 & a_{22} & a_{23} \\ b_3 & a_{32} & a_{33} \end{vmatrix}, \tag{B.6}$$

$$I_2 = \frac{1}{D} \begin{vmatrix} a_{11} & b_1 & a_{13} \\ a_{21} & b_2 & a_{23} \\ a_{31} & b_3 & a_{33} \end{vmatrix}, \tag{B.7}$$

and

$$I_3 = \frac{1}{D} \begin{vmatrix} a_{11} & a_{12} & b_1 \\ a_{21} & a_{22} & b_2 \\ a_{31} & a_{32} & b_3 \end{vmatrix}. \tag{B.8}$$

It remains to evaluate the determinants. For a 3×3 square determinant we have

$$\begin{vmatrix} a & b & c \\ d & e & f \\ g & h & i \end{vmatrix} = aei + bfg + cdh - ceg - bdi - afh. \tag{B.9}$$

There are a few ways to remember this combination. One way is to replicate part of the array in the horizontal direction and then take appropriate diagonal products. For our 3×3 determinant we write

$$\begin{vmatrix} a & b & c & a & b \\ d & e & f & d & e \\ g & h & i & g & h \end{vmatrix}. \tag{B.10}$$

The expression in Eq. (B.9) is then obtained by starting with element a and moving diagonally down and to the right, multiplying the coefficients to obtain aei. The same thing is done with elements b and c to obtain bfg and cdh, respectively. We then do the same thing moving diagonally down and to the left from elements c, a, and b, obtaining ceg, afh, and bdi. These latter products are subtracted from the sum of the former products, giving the desired result. The same method can be used for any size square matrix.

B.3 Using the TI-83

If the actual numeric values of the coefficients a_{ij} and the constants b_i are known, some calculators will perform for you the calculations indicated in Eqs. (B.5) through (B.8). The following summarizes the necessary procedure for the popular TI-83 calculator.

1. Define an $n \times (n + 1)$ matrix, where n is the number of unknown currents. The first n columns contain the coefficients a_{ij} as in Eq. (B.5) and the last column contains the constants b_i. For our three-unknown example, the matrix elements would be

$$\begin{pmatrix} a_{11} & a_{12} & a_{13} & b_1 \\ a_{21} & a_{22} & a_{23} & b_2 \\ a_{31} & a_{32} & a_{33} & b_3 \end{pmatrix}. \tag{B.11}$$

This is accomplished on the TI-83 by hitting MATRIX, moving the cursor over to EDIT, and moving the cursor down to one of the listed matrix names (A, B, C,...). Hit ENTER and you will be prompted for the number of rows and number of columns. Enter this information by moving the cursor and keying in the numbers. Hit ENTER.

2. The cursor now has moved to the first row, first column of the defined matrix display. Enter the appropriate number and hit ENTER. The cursor moves to the next matrix element and you continue to input all the numbers from your problem. When finished, hit QUIT.

3. Hit MATRIX, then select MATH and scroll down to *rref* and hit ENTER. Enter the argument for the *rref* function by hitting MATRIX, selecting names, and scrolling to the name of the matrix you defined in step 1. Hitting ENTER adds this name to the *rref* function.

4. Now hit ENTER once more to run the *rref* function. The result is a new matrix and the last column gives the numeric values of the unknown currents I_1, I_2, etc.

Appendix C: Inductively coupled circuits

C.1 Introduction

Inductance is the expression of Faraday's Law in electronic circuits. Recall that Faraday's Law says that if the magnetic flux through a closed loop changes in time, a voltage will be induced in the loop. In equation form,

$$V = -\frac{d\Phi}{dt}. \tag{C.1}$$

Here, Φ is the magnetic flux through the loop given most generally by

$$\Phi = \int \mathbf{B} \cdot \mathbf{da} \tag{C.2}$$

where the integral is over a surface bounded by the closed loop.

We also know from Ampère's Law that currents produce magnetic fields and those fields encircle the current-carrying wires. Thus, in even the simplest circuit, we have currents producing magnetic fields and those fields producing a magnetic flux through the circuit. If the current is changing in time, then the magnetic field changes in time as does the magnetic flux, giving rise to an induced voltage by Eq. (C.1). Since the magnetic field produced by a current is directly proportional to the current we can say $V \propto -dI/dt$. In the case we have outlined, where the flux through the circuit is caused by the currents in the circuit, the constant of proportionality is called the *self-inductance L* (or simply the *inductance*). The inductance depends on the size, shape, and other geometrical properties of the circuit. In such cases, we include the effect of Faraday's Law in the circuit by writing

$$V = -L\frac{dI}{dt} \tag{C.3}$$

for the voltage across the circuit inductance.

Suppose now that we have two circuits in close proximity. In this case, it is possible for a current I_1 in circuit 1 to produce a magnetic flux not only through circuit 1, but also through circuit 2. A change in I_1 will thus induce a voltage in both circuit 1 and circuit 2. Similarly, a changing current I_2 in circuit 2 will induce a voltage in both circuit 2 and circuit 1. We have already accounted for the voltage induced in a circuit by its own current, but the induced voltage produced by the current in a neighboring circuit requires a new concept, the *mutual inductance M*. The voltage induced in circuit 1 by the variation of current in circuit 2 is given by

$$V_1 = -M\frac{dI_2}{dt}. \tag{C.4}$$

Similarly, the voltage induced in circuit 2 by the variation of current in circuit 1 is given by

$$V_2 = -M\frac{dI_1}{dt}. \tag{C.5}$$

The constant M depends on the size, shape, and other geometrical properties of both of the circuits and is thus a common or mutual property of the circuit pair.

The effect of mutual inductance can be illustrated by considering the circuit pair shown in Fig. C.1. An arbitrary time-varying voltage $V(t)$ drives a series resistor and inductor in circuit 1. An independent, undriven circuit 2 consists of a resistor R_2 and an inductor L_2 in series. The interaction between the two circuits described above is represented by the symbol M.

Applying the voltage loop law to each circuit gives

$$V(t) - I_1 R_1 - L_1\frac{dI_1}{dt} - M\frac{dI_2}{dt} = 0 \tag{C.6}$$

and

$$I_2 R_2 + L_2\frac{dI_2}{dt} + M\frac{dI_1}{dt} = 0. \tag{C.7}$$

Because of the mutual inductance, the current in each circuit depends on the current in the other circuit. We cannot, therefore, solve one equation independently of the other: Eqs. (C.6) and (C.7) are said to be *coupled*.

Since our two equations are linear, we now use our usual complex exponential technique to solve them. Substituting in the complex sinusoidal voltage

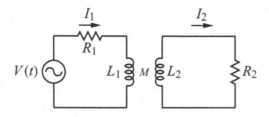

Figure C.1 Example of two circuits coupled by mutual inductance.

$V(t) = V_p e^{j\omega t}$ and currents $I_1 = \hat{I}_{p1} e^{j\omega t}$ and $I_2 = \hat{I}_{p2} e^{j\omega t}$, we obtain[1]

$$V_p - \hat{I}_{p1} R_1 - j\omega L_1 \hat{I}_{p1} - j\omega M \hat{I}_{p2} = 0 \tag{C.8}$$

and

$$\hat{I}_{p2} R_2 + j\omega L_2 \hat{I}_{p2} + j\omega M \hat{I}_{p1} = 0. \tag{C.9}$$

We now have two algebraic equations for two unknowns, $\hat{I}_{p1}$ and $\hat{I}_{p2}$. Solving Eq. (C.9) for $\hat{I}_{p2}$ and plugging into Eq. (C.8) gives

$$\hat{I}_{p1} = V_p \left[\frac{R_2 + j\omega L_2}{\omega^2 (M^2 - L_1 L_2) + R_1 (R_2 + j\omega L_2) + j\omega R_2 L_1} \right]. \tag{C.10}$$

Using this result to eliminate $\hat{I}_{p1}$ from Eq. (C.9) yields

$$\hat{I}_{p2} = \frac{-j\omega M V_p}{\omega^2 (M^2 - L_1 L_2) + R_1 (R_2 + j\omega L_2) + j\omega R_2 L_1}. \tag{C.11}$$

C.2 Transformers

Equations (C.10) and (C.11) are general but not very illuminating. A special case of particular interest occurs for the case of an ideal transformer. In this case, L_1 and L_2 are the inductances of the primary and secondary windings of the transformer. The core of the transformer enhances the coupling between the two circuits by guiding the magnetic field produced by the primary windings through the secondary windings. One can show that the mutual and self-inductances of a transformer are related by $M^2 = k L_1 L_2$, where $0 \leq k \leq 1$. For an ideal transformer (with perfect coupling) $k = 1$ and, thus, $M^2 - L_1 L_2 = 0$. Next, we assume the load resistance is small compared to the impedance of the secondary windings so that R_2 can be ignored compared with $j\omega L_2$. Under these conditions, Eq. (C.10) becomes

$$V_p = \hat{I}_{p1} \left[R_1 + R_2 \frac{L_1}{L_2} \right] \tag{C.12}$$

and Eq. (C.11) becomes

$$V_p = -\hat{I}_{p2} \sqrt{\frac{L_1}{L_2}} \left[R_1 \frac{L_2}{L_1} + R_2 \right]. \tag{C.13}$$

---

[1] The generalization of this method to an arbitrary periodic function is discussed in Section 2.8.

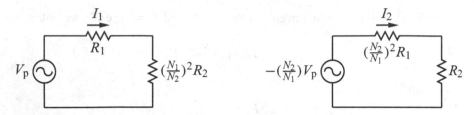

Figure C.2 Equivalent circuits for Eqs. (C.14) and (C.15).

Finally, if we assume the two sets of windings have the same cross-sectional area and the same length (not unreasonable since they are wound on the same core), then $L_1/L_2 = (N_1/N_2)^2$, where N_1 and N_2 are the number of windings on the primary and secondary coils, respectively (see Eq. (2.13)). Equations (C.12) and (C.13) can then be re-written as

$$V_{\mathrm{p}} = \hat{I}_{\mathrm{p}1} \left[R_1 + R_2 \left(\frac{N_1}{N_2} \right)^2 \right] \qquad (\text{C.14})$$

and

$$-\frac{N_2}{N_1} V_{\mathrm{p}} = \hat{I}_{\mathrm{p}2} \left[R_1 \left(\frac{N_2}{N_1} \right)^2 + R_2 \right]. \qquad (\text{C.15})$$

Equations (C.14) and (C.15) describe the two equivalent circuits shown in Fig. C.2. On the right, we see that the secondary circuit is driven by a voltage of magnitude $(N_2/N_1)V_{\mathrm{p}}$. This is in accordance with the first of the basic transformer equations, Eq. (2.126). Solving Eqs. (C.14) and (C.15) for $I_{\mathrm{p}1}$ and $I_{\mathrm{p}2}$ and taking the quotient gives (after some algebra) $I_{\mathrm{p}1}/I_{\mathrm{p}2} = -N_2/N_1$, in accordance with Eq. (2.127). Finally, the circuits show that each resistor affects both circuits, but that the resistor value is transformed: R_2 affects the primary circuit but its value is changed to $(N_1/N_2)^2 R_2$ (this is in accordance with Eq. (2.132)), while R_1 affects the secondary circuit with its value changed to $(N_2/N_1)^2 R_1$.

References

Charles K. Alexander and Matthew N. O. Sadiku, *Fundamentals of Electric Circuits*, 2nd edition (New York: McGraw-Hill, 2004).

L. W. Anderson and W. W. Beeman, *Electric Circuits and Modern Electronics* (New York: Holt, Rinehart, and Winston, 1973).

Dennis Barnaal, *Digital and Microprocessor Electronics for Scientific Application* (Prospect Heights, IL: Waveland Press, 1982).

Dennis Barnaal, *Analog Electronics for Scientific Application* (Prospect Heights, IL: Waveland Press, 1989).

James J. Brophy, *Basic Electronics for Scientists*, 5th edition (New York: McGraw Hill, 1990).

D. V. Bugg, *Electronics: Circuits, Amplifiers and Gates* (New York: Adam Hilger, 1991).

David Casasent, *Electronic Circuits* (New York: Quantum, 1973).

Edwin C. Craig, *Electronics via Waveform Analysis* (New York: Springer, 1993).

A. James Diefenderfer and Brian E. Holton, *Principles of Electronic Instrumentation*, 3rd edition (Philadelphia, PA: Saunders, 1994).

William L. Faissler, *An Introduction to Modern Electronics* (New York: Wiley, 1991).

Earl D. Gates, *Introduction to Electronics*, 5th edition (Clifton Parks, NY: Thomson Delmar Learning, 2007).

Irving M. Gottlieb, *Understanding Oscillators* (Indianapolis, IN: Sams, 1971).

Joseph D. Greenfield, *Microprocessor Handbook* (New York: Wiley, 1985).

Richard J. Higgins, *Electronics with Digital and Analog Integrated Circuits* (Englewood Cliffs, NJ: Prentice-Hall, 1983).

Paul Horowitz and Winfield Hill, *The Art of Electronics*, 2nd edition (New York: Cambridge University Press, 1989).

Richard C. Jaeger and Travis N. Blalock, *Microelectronic Circuit Design*, 3rd edition (New York: McGraw-Hill, 2008).

Walter G. Jung, *IC Timer Cookbook* (Indianapolis, IN: Sams, 1978).

Walter G. Jung, *IC Op-Amp Cookbook*, 3rd edition (Carmel, IN: Sams, 1990).

Charles Kittel, *Introduction to Solid State Physics*, 4th edition (New York: Wiley, 1971).

Don Lancaster, *Micro Cookbook*, Volume 1, *Fundamentals* (Indianapolis, IN: Sams, 1982).

Don Lancaster, *TTL Cookbook* (Indianapolis, IN: Sams, 1983).

Don Lancaster, *Active Filter Cookbook*, 2nd edition (Thatcher, AZ: Synergetics, 1995).

Don Lancaster, *CMOS Cookbook*, 2nd edition (Boston, MA: Newnes, 1997).

Sol Libes, *Small Computer Systems Handbook* (Rochelle Park, NJ: Hayden, 1978).

Donald A. Neamen, *Microelectronics: Circuit Analysis and Design*, 3rd edition (New York: McGraw-Hill, 2007).

Kenneth L. Short, *Microprocessors and Programmed Logic* (Englewood Cliffs, NJ: Prentice-Hall, 1981).

Robert E. Simpson, *Introductory Electronics for Scientists and Engineers*, 2nd edition (Boston, MA: Allyn and Bacon, 1987).

Julien C. Sprott, *Introduction to Modern Electronics* (New York: Wiley, 1981).

Roger L. Tokheim, *Digital Electronics: Principles and Applications*, 6th edition (New York: McGraw-Hill, 2003).

John E. Uffenbeck, *Introduction to Electronics, Devices and Circuits* (Englewood Cliffs, NJ: Prentice-Hall, 1982).

Leopoldo B. Valdes, *The Physical Theory of Transistors* (New York: McGraw-Hill, 1961).

M. Russell Wehr, James A. Richards, Jr., and Thomas W. Adair, III, *Physics of the Atom*, 4th edition (Reading, MA: Addison-Wesley, 1985).

Niklaus Wirth, *Digital Circuit Design for Computer Science Students: An Introductory Textbook* (New York: Springer, 1995).

Index

A/D, *see* analog to digital converter
AC, definition of, 19, 27
ammeter, 13
Ampère's Law, 30, 241
amperes, 1
amplifier
   black box model for, 113
   common-base, 123
   common-collector, 122
   common-drain, 147
   common-emitter, 119
   common-gate, 149
   common-source, 145
   current gain, 113
   distortion, 127
   emitter-follower, 122
   feedback, 128
   frequency response, 127
   input impedance, 113
   open-loop voltage gain, 113
   output impedance, 113
   source-follower, 147
   voltage gain, 113
amplitude
   decibels, 20
   peak, 20
   peak-to-peak, 20
   rms, 20
amplitude modulation, 193
analog to digital converter, 228
anode, 78

band-pass filter, 53
band theory of solids, 69
Barkhausen criterion, 188
battery
   ideal, 3
   real, 23

BCD, *see* binary coded decimal
binary coded decimal, 223
binary counters, 220
binary numbers, 200
bipolar junction transistor, 104
   α, 106
   β, 106
   AC equivalents for, 116
   amplifier circuits, 110
   band structure, 105
   biased for linear active operation, 105
   I–V characteristics, 107
   inverter, 110
   npn, 104
   pnp, 104
   switching circuit, 108
bit, data, 202
BJT, *see* bipolar junction transistor
Boolean algebra, 208
breakpoint frequency, 40

χ, reactance, 48
capacitors, 27
   equivalent circuit laws for, 28
   in parallel, 29
   in series, 28
   voltage rating, 27
carbon, resistivity of, 5
cathode, 78
center-tapped transformer, 87
channel length modulation, 145
charge carriers
   majority, 73
   minority, 73
clamp circuit, 84
clipper circuit, 84
CMOS, 212
complex numbers, 43
   applied to LR circuit, 49

complex numbers (*cont.*)
  applied to LRC circuit, 52
  applied to RC circuit, 45, 48
  complex conjugate of, 45
  magnitude of, 44
  phase of, 44
complex Ohm's Law, 48
conduction band, 71
copper, resistivity of, 5
Cramer's Method, 16, 238
current, definition of, 1
current divider, 12
current limiting, 11
current source, definition of, 11

D/A, *see* digital to analog
     converter
DC, definition of, 19
decoder, 224
DeMorgan's theorems, 209
demultiplexer, 231
determinants, 16, 238
dielectric constant ϵ, 27
digital to analog converter, 227
diode
  $I–V$ characteristic of, 78
  center-tapped full-wave rectifier, 88
  clamp circuit, 84
  clipper circuit, 84
  full-wave bridge rectifier, 90
  half-wave rectifier, 87
  light emitting, 79
  limiter circuit, 84
  logic circuit, 86
  rectifier, 86–90
  simplified model for, 81
  switch protector, 85
  voltage dropper circuit, 83
  zener, 92
doping a semiconductor, 72
duty cycle, 22

energy bands
  definition of, 68
  for a conductor, 69
  for an insulator, 70
  for a semiconductor, 71

energy levels
  atomic, 68
  for a solid, 68
EPROM, 233

farad, 27
Faraday's Law, 30, 241
feedback, 128
FET, *see* field-effect transistor
field-effect transistor, 133
  AC equivalents for, 144
  as a switch, 140
  $I–V$ characteristics for, 136
  junction, 134
  metal oxide semiconductor, 136
    depletion, 136
    enhancement, 136
  model equations for, 136
  pinchoff, 136
  transfer curve for, 140
filters
  band-pass, 53
  high-pass, 40, 51
  low-pass, 41, 50
  power supply, 90
    LC or L-section, 92
    RC π-section, 92
    simple capacitor, 90
555 timer, 180
  astable oscillator, 181
  cascading, 185
  monostable operation, 183
flip-flop
  basic, 216
  binary, 216
  clocked, 217
  gated, 217
  J-K, 219
  master-slave, 218
  M-S, 218
  R-S, 216
forbidden band, 70
Fourier analysis, 58
  sawtooth wave, 60
  square wave, 61
  triangle wave, 61
frequency, 20

frequency domain analysis, 37
frequency modulation, 197
full adder, 214
full-wave bridge rectifier, 90

ground, definition of, 83

h-parameter model, 118
half-adder, 214
half-power frequency, 40
half-wave rectifier, 87
henry, 30
hertz, 20
high-pass filter
    LR, 51
    RC, 40
holes in semiconductors, 72

impedance, 47
    of capacitor, 47
    of inductor, 47
    of resistor, 47
    reactive, 48
    resistive, 48
impedance matching, 63
induced voltage, 30
inductance, 30
    mutual, 242
    self, 241
inductively coupled circuits, 241
inductors
    in parallel, 30
    in series, 30
information registers, 216
input resistance, 17
internal resistance of battery, 23

Karnaugh map, 206
KCL, *see* Kirchoff's Current Law
Kirchoff's Current Law, 2
Kirchoff's Voltage Law, 2
KVL, *see* Kirchoff's Voltage Law

LED, *see* light emitting diode
light emitting diode, 11, 79
limiter circuit, 84
load line method
    applied to BJT switch, 109

applied to FET switch, 140
    for diode circuit, 81
    for zener diode circuit, 93
logic gates, 204–212
    AND, 204
    buffer, 205
    inverter, 206
    making, 211
    NAND, 205
    NOR, 205
    OR, 204
    XNOR, 206
    XOR, 205
low-pass filter
    LR, 50
    RC, 41
LRC circuit, 52
    critically damped response, 58
    frequency response, 53
    overdamped response, 57
    underdamped response, 55

majority charge carriers, 73
matrix, 238
memory chips, 232
mesh loop method, 15
mhos, 144
minority charge carriers, 73
modulo-*n*, 223
multiplexer, 229

nichrome, resistivity of, 5
noise, 22
noise immunity, 200
Norton's theorem, 10
n-type semiconductor, 72

Ohm's Law, 4
ohms, 4
op-amp, *see* operational amplifier
open circuit, definition of, 10
operating point, 81, 112, 116, 141, 144
operational amplifier, 152
    adder, 156
    astable multivibrator, 165
    buffer, 156
    comparator, 153

operational amplifier (*cont.*)
  differential amplifier, 157
  differentiator, 158
  golden rules, 154
  integrator, 158
  inverting amplifier, 155
  inverting input, 152
  non-inverting amplifier, 156
  non-inverting input, 152
  open-loop gain, 153, 164
  practical considerations, 159
    bias currents, 159
    frequency response, 164
    input offset voltage, 162
    slew rate limiting, 162
  saturation voltage, 153
  voltage follower, 156
oscillator
  relaxation, 171
    555 astable, 181
    SCR sawtooth, 171
    transistor astable, 174
  sinusoidal, 185
    crystal, 192
    Hartley, 191
    LC tank circuit, 190
    RC, 186
    stability, 188
    Wein bridge, 189

parallel data transmission, 202
period T, 20
permeability μ, 30
phase, 20
p-n junction
  biased, 76
  breakdown, 78
  depletion region, 74
  energy levels, 74
  forward bias, 77
  photon absorption, 80
  photon emission, 80
  reverse bias, 76
potential difference, 1
potentiometer, 5
power, general definition of, 3
power transfer optimization, 63

prefixes, 3
PROM, 233
p-type semiconductor, 73
pulse train, 22
pulse width, 22

Q point, 112
quiescent point, 112

RAM, 233
ramp, 22
RC circuit, 30–43
  charging, 32
  differentiator, 42
  discharging, 32
  high-pass filter, 40
  integrator, 43
  low-pass filter, 41
  negative phase shifter, 41
  positive phase shifter, 40
  response to sine wave, 37
  response to square wave, 33
RC time constant, 32
reactance χ, 48
rectifier
  diode full-wave, 87, 90
  diode half-wave, 86
  silicon controlled, 97
regulation, 91
regulator
  fixed voltage, 96
  variable voltage, 97
repetition time, 20
resistivity ρ, 5
resistor
  color bands, 5
  current limiting, 11
  equivalent circuit laws for, 6
  $I–V$ characteristic of, 4
  in parallel, 7
  in series, 7
  power laws for, 5
  power rating, 5
  shunt, 13
resonant frequency, 53
rheostat, 5
ringing, 56

ripple factor, 91
roll off, 127
ROM, 233

SCR, *see* silicon controlled rectifier
self-inductance, 30, 241
serial data transmission, 202
seven-segment display, 223
shift register, 224
    digital waveform synthesis, 225
    scrolling display, 224
short circuit, definition, 10
shunt, 13
siemens, 144
silicon controlled rectifier, 97
    as a motor control, 99
    as a switch, 98
    I–V characteristics for, 97
silver, resistivity of, 5
sinusoidal signal, 20
square wave, 21
standard method, 14

thermal energy, 70
thermal transitions, 70
Thevenin's theorem, 10
TI-83, 240
time constant, 32
time domain analysis, 37

transcendental equation, 80
transconductance, 144
transformer, 61, 243
    center tapped, 87
    impedance matching, 63
    primary windings, 62
    secondary windings, 62
    turns ratio, 62
triangle wave, 22
TTL, 212

universal DC bias circuit, 111, 142

valance band, 71
voltage, definition of, 1
voltage divider, 12
voltage dropper circuit, 83
voltage source, definition of, 11
voltmeter, 13
volts, 1

watts, 3
word, data, 232

zener diode, 92
    as regulator, 93
    limiter circuit, 95
    voltage indicator circuit, 95

Printed in the United States
by Baker & Taylor Publisher Services